AF831421

Busy Ant Maths

Test Pack 5

Series Editor: Peter Clarke

Author: Caroline Fawcus

William Collins' dream of knowledge for all began with the publication of his first book in 1819. A self-educated mill worker, he not only enriched millions of lives, but also founded a flourishing publishing house. Today, staying true to this spirit, Collins books are packed with inspiration, innovation and practical expertise. They place you at the centre of a world of possibility and give you exactly what you need to explore it.

Collins. Freedom to teach.

Published by Collins
An imprint of HarperCollins*Publishers*
The News Building
1 London Bridge Street
London
SE1 9GF

HarperCollins*Publishers*
Macken House, 39/40 Mayor Street Upper
Dublin 1, D01 C9W8, Ireland

Browse the complete Collins catalogue at
www.collins.co.uk

10 9 8 7 6 5 4 3 2

ISBN 978-0-00-816740-0

British Library Cataloguing in Publication Data
A catalogue record for this publication is available from the British Library.

Author: Caroline Fawcus
Series Editor: Peter Clarke
Commissioning Editor: Fiona McGlade
Project Editor: Leah Willey
Cover design and artwork: Amparo Barrera
Internal design and illustrations: Steve Evans and Mike Connor
Production: Robin Forrester

Printed and bound in the UK using 100% Renewable Electricity at CPI Group (UK) Ltd

MIX
Paper | Supporting responsible forestry
FSC www.fsc.org
FSC™ C007454

This book is produced from independently certified FSC™ paper to ensure responsible forest management.
For more information visit: www.harpercollins.co.uk/green

Contents

Introduction

Test Papers – Pupil version

Test Papers – Teacher version

Introduction

Purpose of the Busy Ant Maths Year 5 Test Pack

1. To assist teachers in assessing the overall level of mastery achieved by pupils in the mathematics programme of study for Year 5, and in determining pupils' readiness to deal with the expectations of the Year 6 programme of study.
2. To provide parents, teachers and school leaders with information on how pupils are performing in comparison with national standards, by assigning one of the following:
   - working below national standard
   - working towards national standard
   - working at national standard
   - working at greater depth within the national standard.
3. To enable teachers and school leaders to monitor the performance of pupil cohorts, to identify where interventions may be required and to ensure that pupils are supported to achieve sufficient progress and expected attainment.

The Busy Ant Maths Test Packs are intended to be used as summative assessments towards the end of the academic year. They are not designed to assess pupils' level of mastery in each of the individual National Curriculum attainment targets and domains. For guidance on making formative assessments regarding pupils' strengths and weaknesses and level of mastery in specific attainment targets and domains, teachers should refer to the Assessment Tasks and Assessment Exercises in the Busy Ant Maths Assessment Guides.

Structure of the Busy Ant Maths Year 5 Test Pack

<table>
<tr><td>Paper</td><td>Paper 1: arithmetic</td><td>Paper 2: reasoning</td><td>Paper 3: reasoning</td></tr>
<tr><td>Total marks</td><td>40 marks</td><td>35 marks</td><td>35 marks</td></tr>
<tr><td>Recommended timing</td><td>30 minutes</td><td>40 minutes</td><td>40 minutes</td></tr>
<tr><td rowspan="2">National Curriculum domain coverage</td><td colspan="3">Number – Number and place value
Number – Addition and subtraction
Number – Multiplication and division
Number – Fractions (including decimals and percentages)</td></tr>
<tr><td></td><td colspan="2">Measurement
Geometry – Properties of shapes
Geometry – Position and direction
Statistics</td></tr>
</table>

The papers should be administered in order.

Pupils can have a break between the papers.

Administering Paper 1: arithmetic

Recommended timing

Pupils have 30 minutes to complete the paper.

Resources

Pupils will need the following resources:

- a blue or black pen, or a dark pencil
- a rubber (optional). If rubbers are not provided, tell the pupils that they may cross out any answer that they wish to change.

Pupils may also use the following resources, if this is normal classroom practice, provided they only give word-for-word translations:

- bilingual dictionaries or electronic translators
- bilingual word lists
- monolingual English electronic spell checkers.

Pupils are not allowed:

- calculators
- other mathematical equipment, such as angle measurers, mirrors or rulers.

Assistance

Teachers should ensure that nothing they say or do during the test could be interpreted as giving pupils an advantage.

If a pupil requests it, teachers may read a question out loud on a one-to-one basis.

If reading to a pupil, teachers should read words and numbers but not mathematical symbols. This is to ensure that pupils are not given an unfair advantage by having the operation/function inadvertently explained by reading its name.

Before the test begins

- Review the list of pupils with any particular individual needs and ensure that their requirements are met.
- Ensure that pupils are able to work undisturbed and that the classroom layout and seating arrangements are suitable.
- Ensure that any wall displays do not give pupils an unfair advantage.
- Write the start and finish times of the test on the board.
- Provide each pupil with the resources they require.

What to say at the start of the test

- *This is the arithmetic paper.*
- *You will need a blue or black pen or dark pencil. You may use a rubber for this test.* (If rubbers are not provided, tell the pupils that they may cross out any answer that they wish to change.)
- *Write your name and class name/number on the front of your test paper.*

- *Look at the list of instructions on page 1. I will read the instructions to you.* (Read the instructions on page 1 of the test paper to the pupils.)

Instructions

You **may not** use a calculator to answer any questions in this test.

Questions and answers

You have **30 minutes** to complete this test.

Work as quickly and as carefully as you can.

Put your answer in the box for each question.

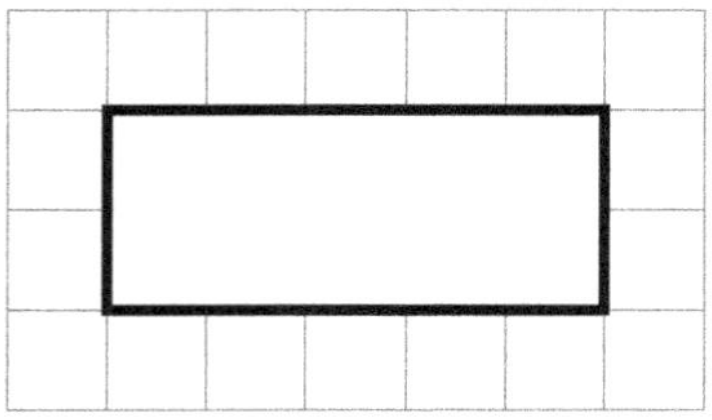

For questions expressed as common fractions, you should give your answer as common fractions.

All other answers should be given as whole numbers or decimals.

If you cannot do one of the questions, **go on to the next one**. You can come back to it later if you have time.

If you finish before the end, **go back and check your work**.

Marks

The number under each box at the side of the page tells you the maximum number of marks for each question.

In this test, some multiplication and division questions are worth 2 marks each. You will be awarded 2 marks for a correct answer. You may get 1 mark for showing a formal method.

All other marks are worth 1 mark each.

- *If you want to change your answer, put a line through the response you don't want, or use a rubber.*
- *Remember to check your work carefully.*
- *If you have any questions during the test, put your hand up and wait for someone to come to you. Remember, I can't help you answer any of the test questions.*
- *You must work on your own and you must not talk to each other.*
- *Are there any questions you want to ask me now?*
- *I will tell you when you have 5 minutes left.*
- *I will tell you when the test is over and to stop writing.*
- *You may now start the test.*

Administering Papers 2 and 3: reasoning

Recommended timing

Pupils have 40 minutes to complete each paper.

Resources

Pupils will need the following resources for each test:

- a blue or black pen, or a dark pencil
- a sharp, dark pencil for mathematical drawing
- a rubber (optional). If rubbers are not provided, tell the pupils that they may cross out any answer that they wish to change.

Pupils may also use the following resources, if this is normal classroom practice, provided they only give word-for-word translations:

- bilingual dictionaries or electronic translators
- bilingual word lists
- monolingual English electronic spell checkers.

Pupils are not allowed:

- calculators.

Assistance

Teachers should ensure that nothing they do or say during the tests could be interpreted as giving pupils an advantage.

If a pupil requests it, teachers may read a question out loud on a one-to-one basis.
If reading to a pupil, teachers should read words and numbers but not mathematical symbols. This is to ensure that pupils are not given an unfair advantage by having the operation/function inadvertently explained by reading its name.
At a pupil's request, teachers can point to parts of the test paper such as charts, diagrams, statements and equations, but they should not explain the information or help the pupil by interpreting it.
If any everyday context or words related to a question are unfamiliar to a pupil, teachers may show them related objects or pictures, or describe the related context.

Before the test begins

- Review the list of pupils with any particular individual needs and ensure that their requirements are met.
- Ensure that pupils are able to work undisturbed and that the classroom layout and seating arrangements are suitable.
- Ensure that any wall displays do not give pupils an unfair advantage.
- Write the start and finish times of the test on the board.
- Provide each pupil with the resources they require.

What to say at the start of the test

- *This is the first/second reasoning paper.*
- *You will need a blue or black pen or dark pencil, and a sharp pencil for mathematical drawing. You may use a rubber for this test.* (If rubbers are not provided, you should tell the pupils that they may cross out any answer that they wish to change.)
- *Write your name and class name/number on the front of your test paper.*

- *Look at the list of instructions on page 1. I will read the instructions to you.* (Read the instructions on page 1 of the test paper to the pupils.)

Instructions

You **may not** use a calculator to answer any questions in this test.

Questions and answers

You have **40 minutes** to complete this test.

Follow the instructions for each question.

Work as quickly and as carefully as you can.

If you need to do working out, you can use the space around the question.
Some questions have a method box like this:

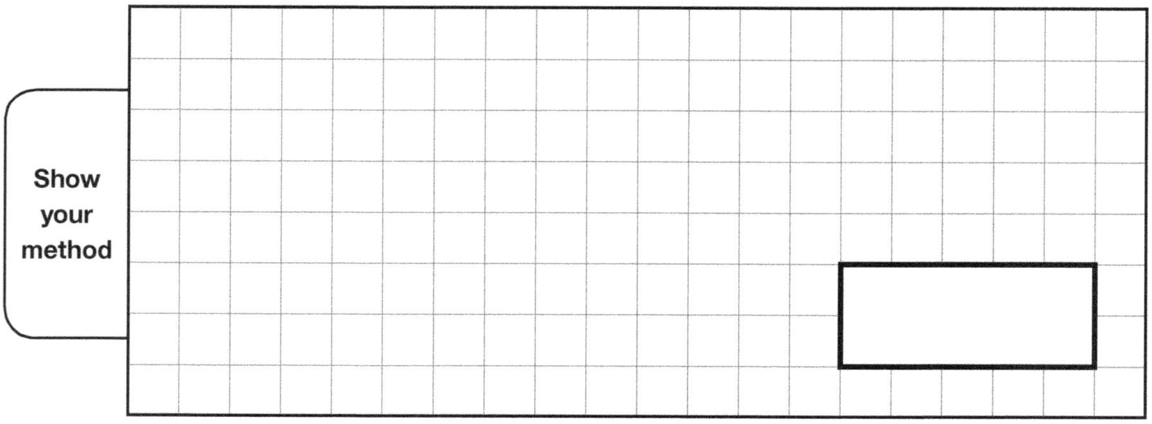

For these questions you may get a mark for showing your method.

If you cannot do one of the questions, **go to the next one.**
You can come back to it later, if you have time.

If you finish before the end, **go back and check your work.**

Marks

The number under each line at the side of the page tells you the maximum number of marks for each question.

- *If you want to change your answer, put a line through the response you don't want, or use a rubber.*
- *Remember to check your work carefully.*
- *If you have any questions during the test, put your hand up and wait for someone to come to you. Remember, I can't help you answer any of the test questions.*
- *You must work on your own and you must not talk to each other.*
- *Are there any questions you want to ask me now?*

- *I will tell you when you have 5 minutes left.*
- *I will tell you when the test is over and to stop writing.*
- *You may now start the test.*

What to do at the end of the test

- Use the mark schemes to mark the three papers, following the '**General marking guidance**', the '**Requirements**', and any '**Additional guidance**' for each question.
- Write the total score for each paper on the front of the paper.
- Calculate the total marks allocated for all three papers to determine the pupil's overall score out of 110. Record their results.
- If you wish to compare pupils' achievements with national standards, refer to the **Assigning a national standards** table at the end of this Introduction.

Explanation of the mark schemes

The marking information for each question is set out in the form of tables which start on page xiii.

The '**Qu**'. column on the left-hand side of each table provides a quick reference to the question number and, where applicable, the question part.

The '**Requirement**' column may include two types of information:

- a statement of the requirements for the award of each mark, with an indication of whether credit can be given for a correct method
- examples of some different types of correct response.

The '**Mark**' column indicates the total number of marks available for each question part.

The '**Additional guidance**' column indicates alternative acceptable responses, and provides details of specific types of response that are unacceptable. Other guidance, such as the range of acceptable answers, is provided as necessary.

General marking guidance

What if...	Marking procedure
The pupil's response is numerically or algebraically equivalent to the answer in the mark scheme.	Award the mark unless the mark scheme states otherwise.
The pupil's response does not match closely any of the examples given.	Teachers should use their professional judgement in deciding whether the response corresponds with the statement of the requirements given in the 'Requirement' column. Consideration should also be given to the 'Additional guidance' column.
The pupil has responded in a non-standard way.	Pupils may provide evidence in a form as long as its meaning can be understood. Diagrams, symbols or words are acceptable for explanations or for indicating a response. In Paper 1: arithmetic, pupils should use formal methods for calculating their answers. For those questions that are worth 2 marks, a partial credit of 1 mark will be awarded for evidence of using formal methods with one arithmetical error. In Papers 2 and 3: reasoning, a partial mark (or marks) will be awarded for evidence of a complete and correct method.
There appears to be a misreading affecting the working.	If the original intention or difficulty level of the question is not reduced deduct 1 mark only. In 1-mark questions, 0 marks are awarded. In 2-mark questions that have a method mark, 1 mark should be awarded if the correct method is correctly implemented with the misread number. If the original intention or difficulty level of the question is reduced, do not award any mark for the question part.
No answer is given in the expected place, but the correct answer is given elsewhere.	Where a pupil has shown understanding of the question, the mark(s) should be given.
The pupil's answer is correct but the wrong working is shown.	A correct response should always be marked as correct unless the mark scheme states otherwise.
The pupil has worked out the answer correctly and then written an incorrect answer in the answer box.	Precedence should be given to the answer given in the answer box over any other workings. There may be cases where the incorrect answer is due to a transcription error. In such cases check the pupil's intention and decide whether to award the mark.
The correct response has been crossed (or rubbed) out and not replaced.	Any legible crossed-out work that has not been replaced should be marked according to the mark schemes. If the work is replaced, then crossed-out work should not be considered.
More than one answer is given.	If all answers are correct (or a range of answers is given, all of which are correct), the mark should be awarded unless prohibited by the mark schemes. If both correct and incorrect responses are given, no mark should be awarded unless the mark scheme states otherwise.
The answer is correct but, in a later part of the question, the pupil has contradicted this response.	A mark given for one part should not be disallowed for working or answers given in a different part, unless the mark scheme specifically states otherwise.

Responses involving money

	Accept	Do not accept
Where the £ is given: £	£1.60 £5 £5.00 Any unambiguous indication of the correct amount, e.g. £1.60p £1 60 pence £1 60 £1,60 £1-60 £1:60	Incorrect placement of pounds or pence, e.g. £160 £160p Incorrect placement of decimal point, or incorrect use or omission of 0, e.g. £1.6 £1 600 £16 0 £1-6-0
Where the p is given: p	20p Any unambiguous indication of the correct amount, e.g. £0.20p	Incorrect or ambiguous use of pounds or pence, e.g. 0.20p £20p
Where no sign is given:	£1.60 160p 20p £0.20 Any unambiguous indication of the correct amount, e.g. £1.60p £1 60 pence £1 60 £1,60 £1-60 £1:60 1.60 160 1 pound 60 £0.20p £.20p £.20 20 0.20	Incorrect or ambiguous use of pounds or pence, e.g. £160 £160p £1.6 1.60p £20 £20p 0.2 0.20p

Responses involving time

	Accept	Do not accept
A time interval, e.g. 1 hour 30 minutes	1 hour 30 minutes Any unambiguous, correct indication, e.g. $1\frac{1}{2}$ hours 1·5 hours 1h 30 1h 30 min 1 30 90 minutes 90 Digital electronic time, i.e. 1:30	Incorrect or ambiguous time interval, e.g. 1.30 1-30 1,30 130 1·3 1.3 hours 1·3h 1h 3 1·30 min
A specific time, e.g. 3:20 a.m., 15:20	3:20 a.m. 3:20 twenty past three Any unambiguous, correct indication, e.g. 03.20 3·20 0320 3 20 3-20 3,20 Unambiguous change to 12- or 24-hour clock, e.g.: 18:10 as 6:10 p.m. or 18:10 p.m.	Incorrect time, e.g. 3·2 a.m. 3·20 p.m. Incorrect placement of separators, spaces, etc. or incorrect use or omission of 0, e.g. 320 3:2:0 3·2 032

Responses involving measures

	Accept	Do not accept
Where units are given, e.g. m, kg, l m	1·5 m Any unambiguous indication of the correct measurement, e.g. $1\frac{1}{2}$ m 1·50 m 1 m 50 cm	Incorrect or ambiguous use of units, e.g. 1500 m

Notes:

- If a pupil leaves the answer box empty but writes the answer elsewhere on the page, then that answer must be consistent with the units given in the answer box and the conditions listed above.
- If a pupil changes the unit given in the answer box, then their answer must be equivalent to the correct answer using the unit they have chosen, unless otherwise indicated in the mark schemes.

Mark schemes for Paper 1: arithmetic

Qu.	Requirement	Mark	Additional guidance
1	1800	1	
2	10 112	1	
3	$\frac{2}{7}$	1	Accept equivalent fractions or the exact decimal equivalent, e.g. $0.\overline{285714}$ Do not accept rounded or truncated decimals.
4	121	1	
5	11 862	1	
6	270	1	
7	4·156	1	
8	$4\frac{1}{2}$	1	Accept equivalent fractions or decimals.
9	£91.26	1	
10	1278	1	
11	156	1	
12	9	1	
13	22 203	1	
14	1	1	
15	21	1	
16	125	1	
17	$\frac{3}{14}$	1	
18	39 608	1	
19	Award **TWO** marks for the correct answer of 88 If the answer is incorrect, award **ONE** mark for a formal short division method which contains no more than **ONE** arithmetical error, e.g. Wrong answer $4\overline{)3\ \ 5\ \ {}^{3}2}$	**Up to 2**	Short division methods must be supported by evidence of appropriate carrying figures to indicate the use of a division algorithm, and must be a complete method.

Qu.	Requirement	Mark	Additional guidance
20	$\frac{7}{10}$	1	Accept equivalent fractions or the exact decimal equivalent, e.g. 0·7
21	27	1	
22	2400	1	
Question 22 commentary: Pupils are expected to use their knowledge of number facts to answer this question.			
23	7518	1	
24	Award **TWO** marks for the correct answer of 6057 If the answer is incorrect, award **ONE** mark for a formal multiplication method which contains no more than **ONE** arithmetical error, e.g. 2019 × 3 Wrong answer 2	Up to 2	Short multiplication methods must be supported by evidence of appropriate carrying figures to indicate the use of a multiplication algorithm, and must be a complete method. **Do not** award any method marks for a place value error, e.g. 2019 × 3 657 2
25	17·8	1	Accept equivalent fractions, e.g. $17\frac{4}{5}$
26	17·181	1	
27	600	1	
28	100 999	1	
29	8	1	
30	14·75	1	**Do not** accept 14 remainder 3 Accept $14\frac{3}{4}$
31	20	1	

Qu.	Requirement	Mark	Additional guidance
32	Award **TWO** marks for the correct answer of 2244 If the answer is incorrect, award **ONE** mark for a formal multiplication method which contains no more than **ONE** arithmetical error, e.g. 132 × 17 914 1320 Wrong answer	Up to 2	**Do not** award any marks if: • the error is in the place value, e.g. the omission of the zero when multiplying by tens: 132 × 17 924 132 Wrong answer • the final (answer) line of the digits is missing. Working must be carried through to reach an answer for the award of **ONE** mark.
33	70	1	
34	Award **TWO** marks for the correct answer of 153 If the answer is incorrect, award **ONE** mark for a formal short division method which contains no more than **ONE** arithmetical error, e.g. Wrong answer $7\overline{)1\;0\;{}^{3}7\;{}^{2}1}$	Up to 2	Short division methods must be supported by evidence of appropriate carrying figures to indicate the use of a division algorithm, and must be a complete method.
35	10	1	
36	21	1	

Mark schemes for Paper 2: reasoning

<table>
<tr><th>Qu.</th><th>Requirement</th><th>Mark</th><th>Additional guidance</th></tr>
<tr><td>1</td><td>800 000 circled, as shown:
80 000   (800 000)   8 000 000   80 000 000</td><td>1</td><td></td></tr>
<tr><td>2</td><td>3415 km</td><td>1</td><td></td></tr>
<tr><td>3</td><td>A correctly completed diagram, as shown:
28 → 14 → 7 → 3·5 → 1·75</td><td>1</td><td></td></tr>
<tr><td>4</td><td>Award TWO marks for a correctly completed Carroll diagram, as shown:
<table>
<tr><th></th><th>multiple of 6</th><th>not a multiple of 6</th></tr>
<tr><td>square number</td><td>36</td><td>16</td></tr>
<tr><td>not a square number</td><td>24, 30</td><td>27</td></tr>
</table>
If the answer is incorrect, award ONE mark for any two numbers correctly placed.</td><td>Up to 2</td><td></td></tr>
<tr><td>5</td><td>–8 °C</td><td>1</td><td>Do not accept 8– °C</td></tr>
<tr><td>6</td><td>DC</td><td>1</td><td>Do not accept 600</td></tr>
<tr><td>7</td><td>Award TWO marks for a correctly completed table, as shown:
<table>
<tr><th>Number</th><th>Factors</th></tr>
<tr><td>8</td><td>2, 4</td></tr>
<tr><td>28</td><td>2, 4, 7, 14</td></tr>
<tr><td>40</td><td>2, 4, 5, 8, 10, 20</td></tr>
</table>
If the answer is incorrect, award ONE mark for any one row completed correctly.</td><td>Up to 2</td><td>Do award the mark if 1 and 28 are listed as factors.
Do not award the mark if any extra numbers are given.</td></tr>
<tr><td>8</td><td>Any correctly completed calculation, e.g.
55 + 8 = 63
55 + 17 = 72
55 + 26 = 81
55 + 35 = 90
55 + –1 = 54</td><td>1</td><td>Award the mark for correct fractional or negative numbers.
Do not award the mark if any box is left blank, e.g.
55 + 8 = ☐</td></tr>
<tr><td>9</td><td>Fractions placed in correct order, as shown:
$\frac{15}{5}$ $\frac{17}{5}$ 3·5 $3\frac{3}{5}$
smallest   largest</td><td>1</td><td>Award the mark if equivalent values are given in answer box, e.g.
3, 3·4, 3·5, 3·6</td></tr>
<tr><td>10</td><td>64</td><td>1</td><td></td></tr>
</table>

<table>
<tr><th>Qu.</th><th>Requirement</th><th>Mark</th><th>Additional guidance</th></tr>
<tr><td>11</td><td>Award TWO marks for a correct answer of $\frac{4}{9}$
If the answer is incorrect, award ONE mark for evidence of a correct method with no more than ONE arithmetical error, e.g.
$\frac{2}{9} + \frac{1}{3} = \frac{5}{9}$
$1 - \frac{5}{9}$ = wrong answer
OR
$1 - \frac{2}{9} = \frac{7}{9}$
$\frac{7}{9} - \frac{1}{3}$ = wrong answer</td><td>Up to 2</td><td>Accept an equivalent fraction or an exact decimal, i.e. $0{\cdot}\dot{4}$</td></tr>
<tr><td>12</td><td>36 599</td><td>1</td><td></td></tr>
<tr><td>13</td><td>34</td><td>1</td><td></td></tr>
<tr><td>14a
14b</td><td>290°
reflex</td><td>1
1</td><td></td></tr>
<tr><td>15</td><td>$\frac{47}{100}$</td><td>1</td><td>Accept any equivalent fraction.</td></tr>
<tr><td>16</td><td>90 cm ticked, as shown:
9 cm ☐
30 cm ☐
90 cm ☑ ✔
3 m ☐
9 m ☐</td><td>1</td><td></td></tr>
<tr><td>17</td><td>Both rows completed correctly, as shown:
<table>
<tr><th></th><th>Less than $\frac{1}{2}$</th><th>Greater than $\frac{1}{2}$</th></tr>
<tr><td>$\frac{3}{4}$</td><td></td><td>✔</td></tr>
<tr><td>$\frac{11}{20}$</td><td></td><td>✔</td></tr>
<tr><td>$\frac{42}{100}$</td><td>✔</td><td></td></tr>
</table></td><td>1</td><td></td></tr>
<tr><td>18</td><td>Award TWO marks for a correct answer of 1170 g or 1·170 kg
If the answer is incorrect, award ONE mark for a correct method, with no more than ONE arithmetical error, e.g.
130 × 8 = 1040
1040 + 130 = wrong answer
OR
130 × 9 = wrong answer</td><td>Up to 2</td><td>For the award of TWO marks, units must be given with the answer.
Award TWO marks for a correct answer that is subsequently spoiled, e.g.
1170 g = 11·7 kg (incorrect)</td></tr>
<tr><td>19</td><td>0·321</td><td>1</td><td></td></tr>
</table>

Qu.	Requirement	Mark	Additional guidance
20	Isosceles **and** a correct explanation. The explanation must state that the sides of a regular pentagon are equal **and** an isosceles triangle has two equal sides, e.g. • 'The pentagon is regular and so two sides of the triangle are equal.'	1	**Do not** award the mark for stating that an isosceles triangle has two equal sides without reference to the properties of a regular pentagon. Explanations that are **not acceptable** include: • The pentagon is regular. • An isosceles triangle has two equal sides. • The pentagon has equal sides.
21	A line joining 70 × 25 and 70 × 5, as shown: 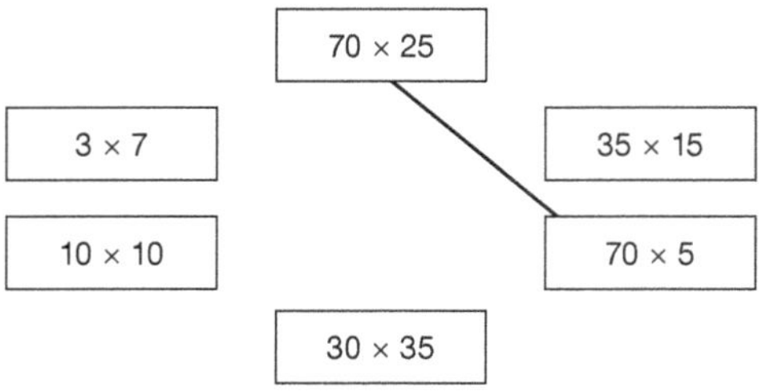	1	
22a 22b	1124 123 minutes	1 1	
23	The calculation 604 + 97 correctly identified. 703 − 99 ☐ 5·2 × 104 ☐ 1197 ÷ 3 ☐ 5000 ÷ 504 ☐ 604 + 97 ☑ 32 × 9 ☐	1	Pupils are expected to be able to make estimations. In this question, it is not expected that answers are worked out exactly.
24	Award **TWO** marks for a correct answer of 32 cm If the answer is incorrect, award **ONE** mark for a method that shows the addition of six sides, e.g. 7 + 4 + 3 + 5 + 4 + 9 = wrong answer	**Up to 2**	
25	An explanation which includes that 63 has factors other than 1 and itself, e.g. 9 × 7 OR 7 is a factor.	1	**Do not** award the mark if either 7 or 9 have not been identified, e.g. 'A prime number has no factors, and 63 does.' is not sufficient for the mark.
26	20%	1	
27	Award **TWO** marks for a correct answer of 26 If the answer is incorrect, award **ONE** mark for a complete correct method with no more than **ONE** arithmetical error, e.g. 1 m ÷ 50 cm = 2 6·5 m ÷ 50 cm = 13 2 × 13 = wrong answer	**Up to 2**	

Mark schemes for Paper 3: reasoning

Qu.	Requirement	Mark	Additional guidance
1	The rectangle ticked, as shown: 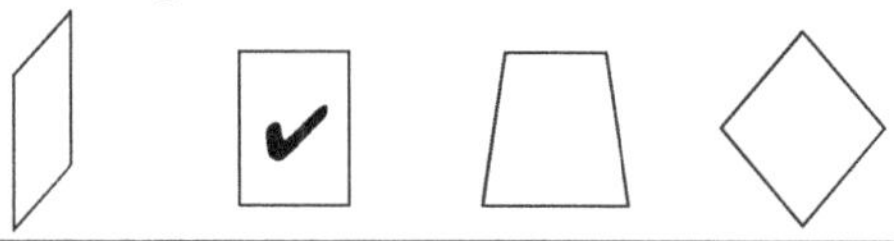	1	The pupil should tick the shape, but an unambiguous indication of the correct answer will gain the mark.
2	700 000	1	
3	606 066	1	
4a 4b	teddy and guitar 95p **OR** £0.95	1 1	Answers can be given in either order.

Question 4 commentary: The second part of the question does not specify the units to be used in the answer and so any answers given in either pounds or pence will be awarded the mark. See the 'General marking guidance' section for further guidance on acceptable answers.

Qu.	Requirement	Mark	Additional guidance
5a 5b	538 639 3 919 718	1 1	
6	All lengths given in correct order, i.e. 3 m 7 cm, 3·4 m, $3\frac{1}{2}$ m, 360 cm smallest … largest	1	Award the mark if equivalent lengths are given in the answer box, e.g. 307 cm, 3·4 m, 350 cm, 360 cm
7a 7b	4 cm 1 cm	1 1	 Accept answers in the range 0·75–1·25 cm.

Question 7 commentary: In the second part of the question, pupils are required to read information from the graph in order to answer the question. Since the points are in between two grid lines, pupils are only able to estimate the exact values.

Qu.	Requirement	Mark	Additional guidance
8	3, 5, 7, 11	1	Award the mark if the correct prime numbers are not given in sequence.
9	A two-part instruction, given in either order: RIGHT 2 **AND** DOWN 3	1	**Do not** accept instructions of more than two parts, e.g. RIGHT 3, LEFT 1, DOWN 3 **Do not** accept arrows (→←) instead of RIGHT/LEFT.
10a 10b	5400 A single instruction to multiply by 1000, e.g. 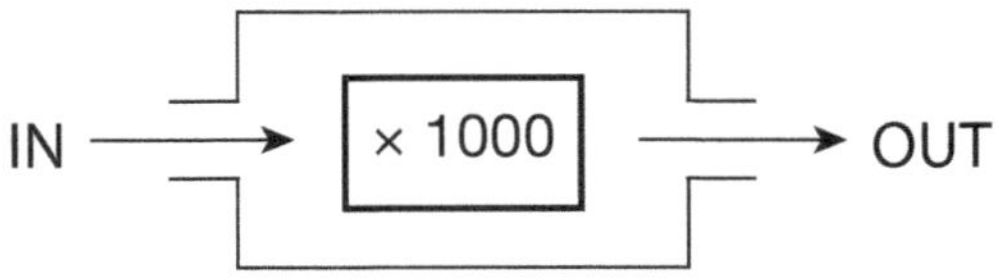	1 1	 Award the mark if the instruction is written in words, e.g. multiply by 1000

<table>
<tr><th>Qu.</th><th>Requirement</th><th>Mark</th><th>Additional guidance</th></tr>
<tr><td>11</td><td>Award TWO marks for all three fractions (excluding the example from the question) correctly linked to the number line, as shown:
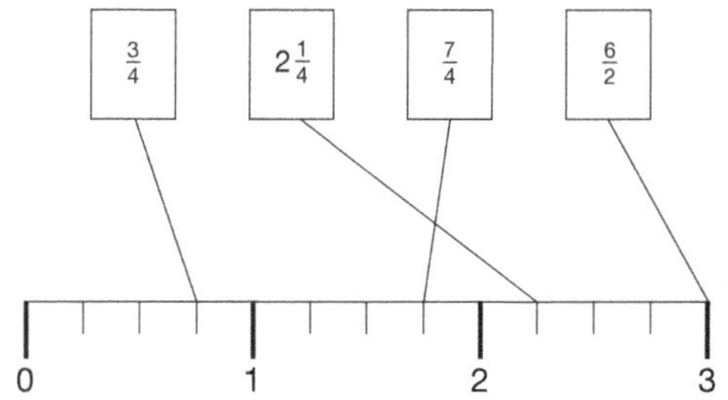

If the answer is incorrect, award ONE mark for any TWO of the three answers that are correct.</td><td>Up to 2</td><td></td></tr>
<tr><td>12</td><td>8400</td><td>1</td><td></td></tr>
<tr><td>13</td><td>Award ONE mark for each correct answer in the correct position of the table, as shown:
<table><tr><th></th><th>Rounded to a whole number</th><th>Rounded to one decimal place</th></tr><tr><td>2·14</td><td>2</td><td>2·1</td></tr><tr><td>5·67</td><td>6</td><td>5·7</td></tr></table></td><td>Up to 2</td><td></td></tr>
<tr><td>14</td><td>Award TWO marks for all three lengths correctly given, as shown:
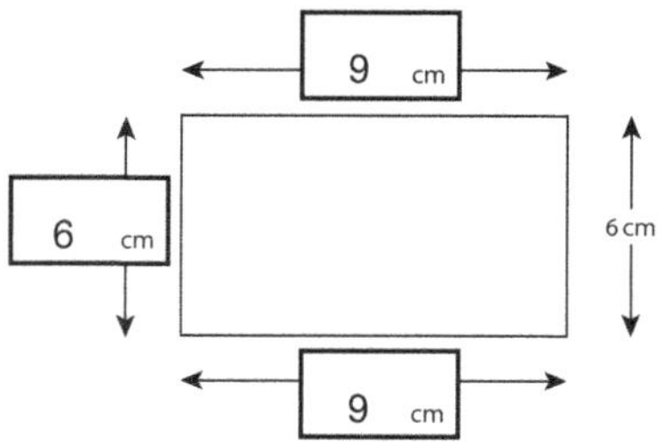

If the answer is incorrect, award ONE mark for any TWO lengths correctly given.</td><td>Up to 2</td><td></td></tr>
<tr><td>15</td><td>Award TWO marks for all four pairs correctly matched, as shown:
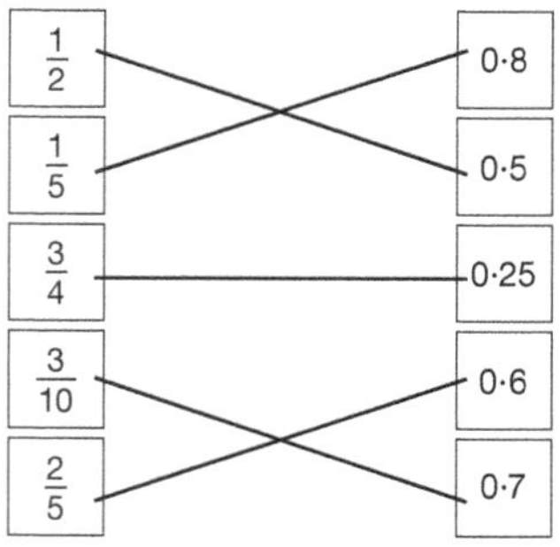

If the answer is incorrect, award ONE mark for any TWO pairs correctly matched.</td><td>Up to 2</td><td>Mappings must be a one-to-one mapping to be considered correct.</td></tr>
</table>

Qu.	Requirement	Mark	Additional guidance
16a 16b	100 86	1 1	
17	Award **TWO** marks for the correct answer of 125 ml. If the answer is incorrect, award **ONE** mark for evidence of a correct method, e.g. 1000 ÷ 8 = OR 1 ÷ 8 = 0·125 0·125 × 1000	Up to 2	Award **ONE** mark for an answer of 0·125 ml. A final answer need not be given for the award of **ONE** mark.
18	B and D	1	Answers may be given in either order.
19	7486 circled, as shown: 8200 (7486) 8473 7745 8005	1	
20	Award **TWO** marks for the correct answer of 84 If the answer is incorrect, award **ONE** mark for evidence of a correct method with no more than **ONE** arithmetical error, e.g. 15 × 12 = 180 32 × 3 = 96 180 – 96 = wrong answer	Up to 2	
21	$\frac{15}{6}$	1	Accept any equivalent fraction, e.g. $\frac{5}{2}$, $2\frac{3}{6}$ or $2\frac{1}{2}$
22	3·5	1	**Do not** accept $3\frac{1}{2}$
23	Award **TWO** marks for all four shapes correctly matched, as shown: 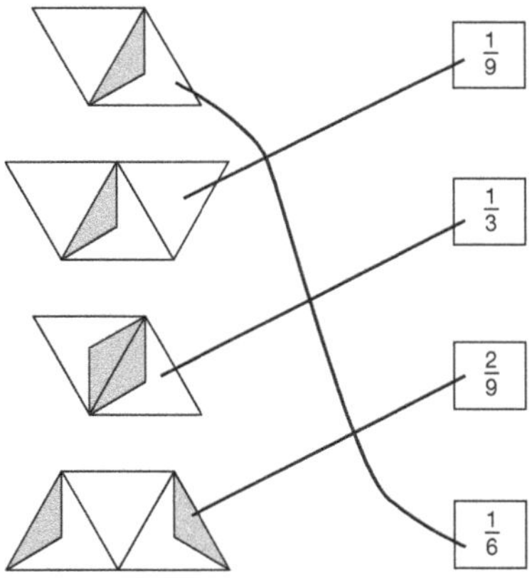 If the answer is incorrect, award **ONE** mark for any **TWO** correctly matched.	Up to 2	

Assigning a national standard

A total of 110 marks are available for the Year 5 Test:

- Paper 1: arithmetic (40 marks)
- Paper 2: reasoning (35 marks)
- Paper 3: reasoning (35 marks)

The sum of the marks allocated from these two papers determines the pupil's overall raw score.

Schools that wish to compare their pupils' achievements with national standards should use the thresholds in the table below to convert raw scores into national standards.

Raw score (out of 110)	0–21	22–54	55–87	88–110
Scaled score	0%–19%	20%–49%	50%–79%	80%–100%
Standard	Working below national standard	Working towards national standard	Working at national standard	Working at greater depth within the national standard

Year 5
PAPER
1

Arithmetic

Name: ____________________

Class: ____________________

Instructions

You **may not** use a calculator to answer any questions in this test.

Questions and answers

You have **30 minutes** to complete this test.

Work as quickly and as carefully as you can.

Put your answer in the box for each question.

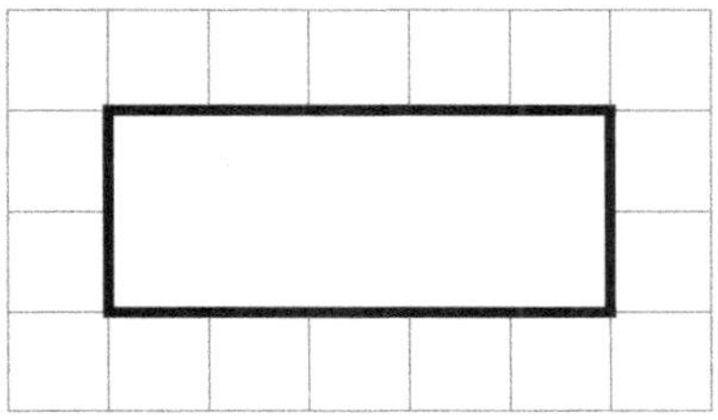

For questions expressed as common fractions, you should give your answer as common fractions.

All other answers should be given as whole numbers or decimals.

If you cannot do one of the questions, **go on to the next one**. You can come back to it later if you have time.

If you finish before the end, **go back and check your work**.

Marks

The number under each box at the side of the page tells you the maximum number of marks for each question.

In this test, some multiplication and division questions are worth 2 marks each. You will be awarded 2 marks for a correct answer. You may get 1 mark for showing a formal method.

All other marks are worth 1 mark each.

1

$18 \times 100 =$

1 mark

2

$10\,000 + 112 =$

1 mark

3

$\frac{5}{7} - \frac{3}{7} =$

1 mark

4 $11^2 =$

1 mark

5 $8071 + 3791 =$

1 mark

6 $27\ 000 \div 100 =$

1 mark

7 $1{\cdot}047 + 3{\cdot}109 =$

1 mark

8 $\frac{1}{2} \times 9 =$

1 mark

9 £100 – £8.74 =

1 mark

10 $213 \times 6 =$

1 mark

11 $13 \times 12 =$

1 mark

12 10% of $90 =$

1 mark

13 $20\,001 + 2001 + 201 =$

1 mark

14 $0{\cdot}65 + 0{\cdot}35 =$

1 mark

15 $168 \div 8 =$

1 mark

16 $5 \times 5 \times 5 =$

1 mark

17 $\frac{1}{7} + \frac{1}{14} =$

1 mark

18 $40\,000 - 392 =$

1 mark

19

$$4\overline{)3\ 5\ 2}$$

Show your method

2 marks

20

$$\frac{2}{5} + \frac{3}{10} =$$

1 mark

21

$$3^3 =$$

1 mark

22 $48 \times 50 =$

1 mark

23 $15\,937 - 8419 =$

1 mark

24

	2	0	1	9
×				3

Show your method

2 marks

25

$178 \div 10 =$

1 mark

26

$14{\cdot}61 + 2{\cdot}571 =$

1 mark

27

$2 \times 6 \times 50 =$

1 mark

28 $99\,999 + 1000 =$

1 mark

29 $\frac{4}{5} \times 10 =$

1 mark

30 $59 \div 4 =$

1 mark

31 $300 \div 15 =$

1 mark

32

$$\begin{array}{r} 132 \\ \times \quad 17 \\ \hline \end{array}$$

Show your method

2 marks

33 $0{\cdot}07 \times 1000 =$

1 mark

34

$$7\overline{)1071}$$

Show your method

2 marks

35

25% of 40 =

1 mark

36

$15 \times 1\frac{2}{5} =$

1 mark

Year 5
PAPER
2

Reasoning

Name: ____________________

Class: ____________________

Instructions

You **may not** use a calculator to answer any questions in this test.

Questions and answers

You have **40 minutes** to complete this test.

Follow the instructions for each question.

Work as quickly and as carefully as you can.

If you need to do working out, you can use the space around the question.

Some questions have a method box like this:

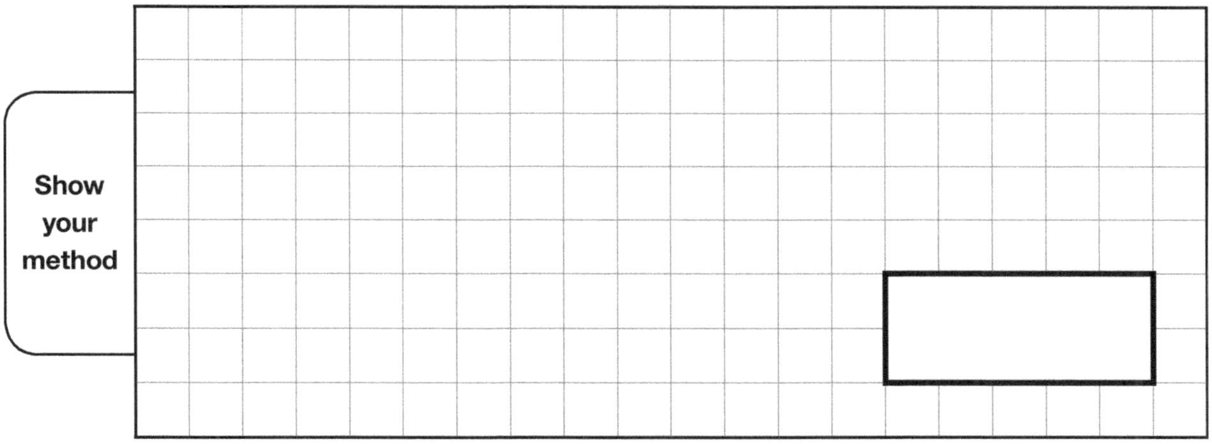

For these questions you may get a mark for showing your method.

If you cannot do one of the questions, **go to the next one.**
You can come back to it later, if you have time.

If you finish before the end, **go back and check your work.**

Marks

The number under each line at the side of the page tells you the maximum number of marks for each question.

1 Circle the number that is between half a million and one million.

80 000 800 000 8 000 000 80 000 000

1 mark

2 The chart shows the distance in kilometres between some cities in Australia.

Alice Springs						
3012	Brisbane					
2324	1717	Cairns				
1511	3415	2727	Darwin			
2270	1674	3054	3781	Melbourne		
3630	4384	5954	4045	3452	Perth	
2644	996	2546	4000	868	4144	Sydney

What is the distance between Brisbane and Darwin?

km

1 mark

3 The rule for a number sequence is to **halve** each time.

Complete the number sequence.

28 → 14 → 7 → ☐ → ☐

1 mark

4 Write the numbers in the Carroll diagram.

One has been done for you.

~~16~~ 24 27 30 36

	multiple of 6	**not a multiple of 6**
square number		16
not a square number		

2 marks

5 The table shows the minimum temperatures in five different cities on one day.

Boston	7 °C
Chicago	–2 °C
Detroit	4 °C
Buffalo	–6 °C
Denver	°C

The minimum temperature in Denver was 6 degrees colder than in Chicago.

Complete the table.

1 mark

6 **Write the missing Roman numeral in the box to complete the sequence.**

DXCVIII, DXCIX, ☐, DCI

1 mark

7 The table shows the factors of a number, excluding 1 and itself.

Write the missing numbers in the table.

Number	Factors
8	2, 4
28	
	2, 4, 5, 8, 10, 20

2 marks

8 Complete the calculation so that the answer is a multiple of 9

55 + ☐ = ☐

1 mark

9 Write the fractions and decimal in order of size, starting with the smallest.

$3\frac{3}{5}$ $\frac{15}{5}$ $3{\cdot}5$ $\frac{17}{5}$

smallest largest

1 mark

10 Write the **cube number** between 50 and 100

1 mark

11 Alicia, Kayden and Joe paint a wall for the school play.

Alicia paints $\frac{2}{9}$ of the wall.

Kayden paints $\frac{1}{3}$ of the wall.

What fraction of the wall does Joe paint?

Show your method

2 marks

12 Write the missing number in the box.

13 401 + [ ] = 50 000

1 mark

13 One crate holds 6 bottles.

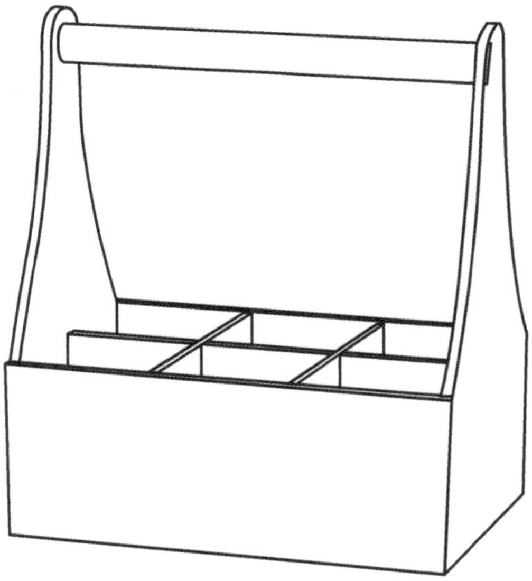

How many crates are needed to hold 204 bottles?

[ ]

1 mark

14 Calculate the size of angle *a*

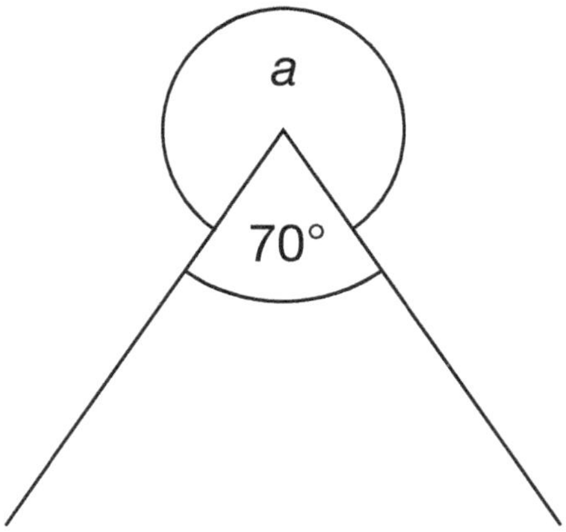

°

1 mark

Complete the table.

	type of angle
angle ***a***	
angle of 70°	acute

1 mark

15 Write 0·47 as a fraction.

1 mark

16 A pond is approximately 3 feet deep.

Tick (✔) the measure that is approximately 3 feet in metric units.

9 cm ☐

30 cm ☐

90 cm ☐

3 m ☐

9 m ☐

1 mark

17 Complete the table.

The first row has been done for you.

	Less than $\frac{1}{2}$	Greater than $\frac{1}{2}$
$\frac{3}{4}$		✔
$\frac{11}{20}$		
$\frac{42}{100}$		

1 mark

18 The mass of an orange is 130 g

A pineapple is **8 times heavier**.

Give units with your answer.

Calculate the total mass of the orange and the pineapple.

Show your method

2 marks

19 Here are five cards.

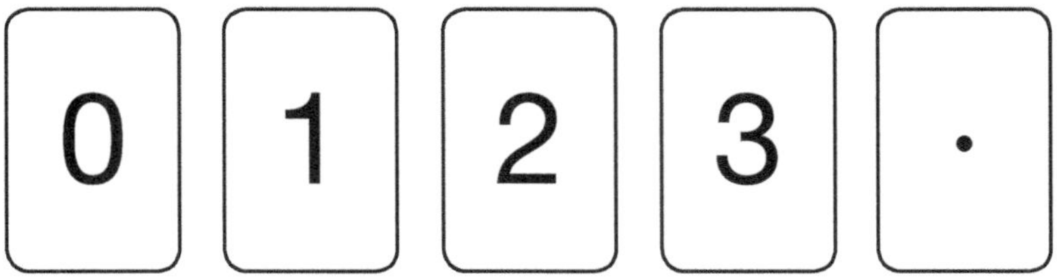

Use each card once to make the largest number less than 1

1 mark

20 A regular pentagon is cut to create a trapezium and a triangle.

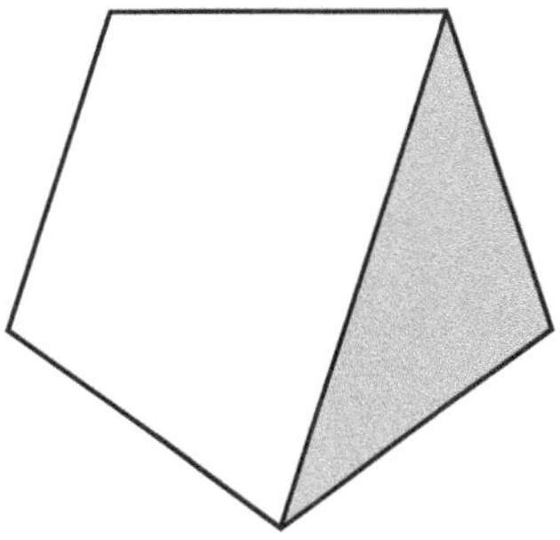

Tick (✔) the type of triangle.

isosceles ☐ equilateral ☐ scalene ☐

Explain your answer.

____

____

1 mark

21 Here is a number fact:

$$70 \times 30 = 2100$$

Draw a line to join **two** cards that have a total of 2100

	70×25	
3×7		35×15
10×10		70×5
	30×35	

1 mark

22 Here is a bus timetable for the town of Thriplow.

Bus timetable

Station	0917	1107	1314	1515	1714
Park	0934	1124	1331	1552	1751
Market	1015	1205	1412	1613	1812
Library	1057	1247	1454	1655	1854
Airport	1137	1327	1534	1735	1934

Jake is at the park at a **quarter to eleven**.

What time does the next bus depart?

1 mark

The bus takes 2 hours and 3 minutes to travel from the park to the airport.

Write this time in minutes.

minutes

1 mark

23 Here are six calculations.

Tick (✔) the calculation that gives the **largest** answer.

$703 - 99$	☐	$5{\cdot}2 \times 104$	☐
$1197 \div 3$	☐	$5000 \div 504$	☐
$604 + 97$	☐	32×9	☐

1 mark

24 Find the perimeter of the shape.

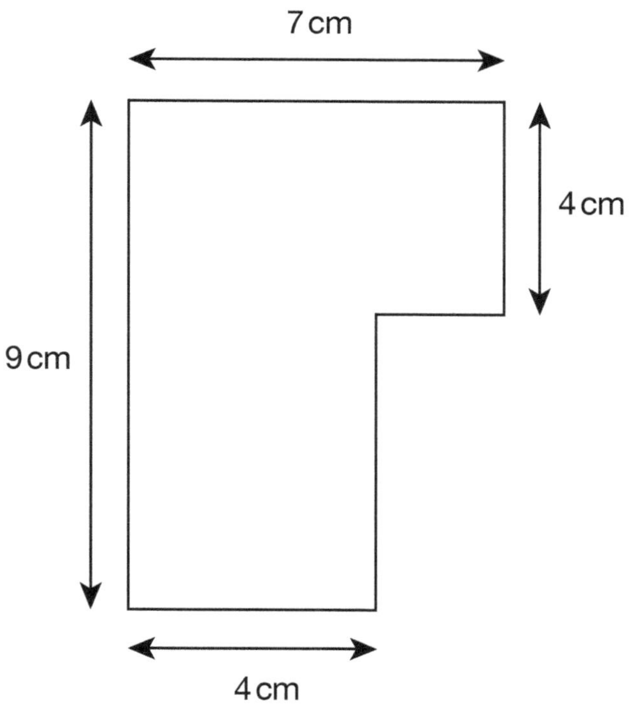

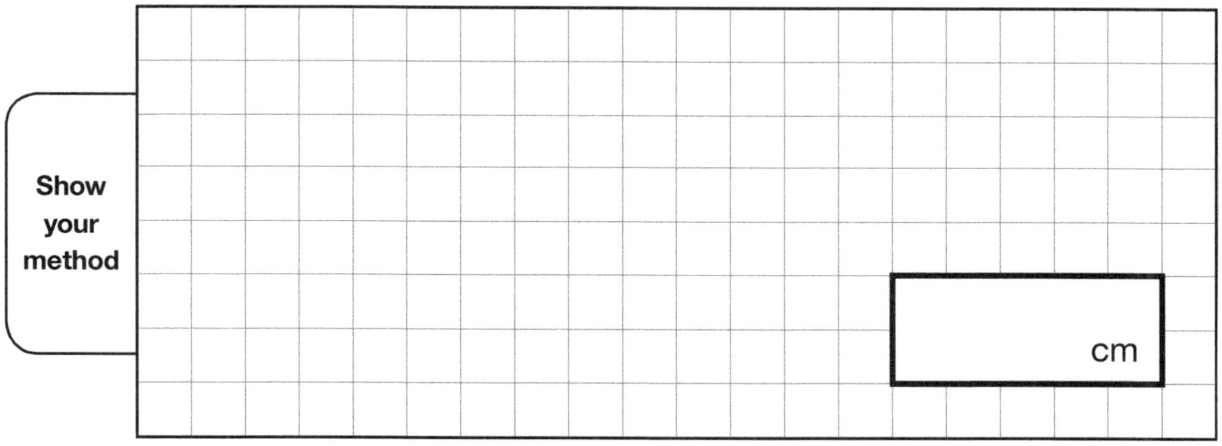

2 marks

25 Francesca says:

Show that she is wrong, using an example.

____

____ 1 mark

26 The multiples of 5 are shaded on the 100 square.

1	2	3	4	5	6	7	8	9	10
11	12	13	14	15	16	17	18	19	20
21	22	23	24	25	26	27	28	29	30
31	32	33	34	35	36	37	38	39	40
41	42	43	44	45	46	47	48	49	50
51	52	53	54	55	56	57	58	59	60
61	62	63	64	65	66	67	68	69	70
71	72	73	74	75	76	77	78	79	80
81	82	83	84	85	86	87	88	89	90
91	92	93	94	95	96	97	98	99	100

What **percentage** of the numbers from 1 to 100 are multiples of 5?

%

1 mark

27 A piece of material measures 6·5 m by 1 m

How many 50 cm squares can be cut from the material?

Show your method

2 marks

Year 5
PAPER
3

Reasoning

Name: ______________

Class: ______________

Instructions

You **may not** use a calculator to answer any questions in this test.

Questions and answers

You have **40 minutes** to complete this test.

Follow the instructions for each question.

Work as quickly and as carefully as you can.

If you need to do working out, you can use the space around the question.

Some questions have a method box like this:

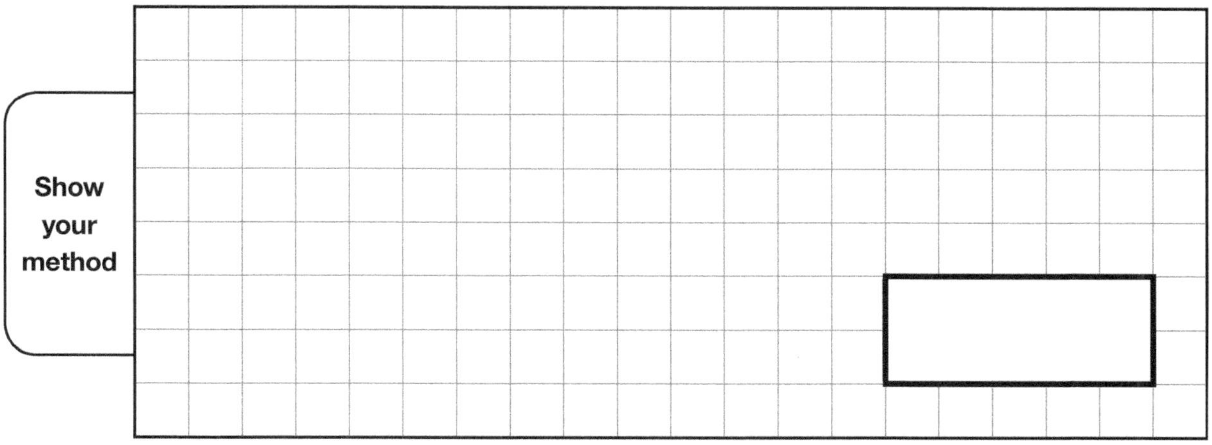

For these questions you may get a mark for showing your method.

If you cannot do one of the questions, **go to the next one.**
You can come back to it later, if you have time.

If you finish before the end, **go back and check your work.**

Marks

The number under each line at the side of the page tells you the maximum number of marks for each question.

1 One face of the 3-D shape is shaded.

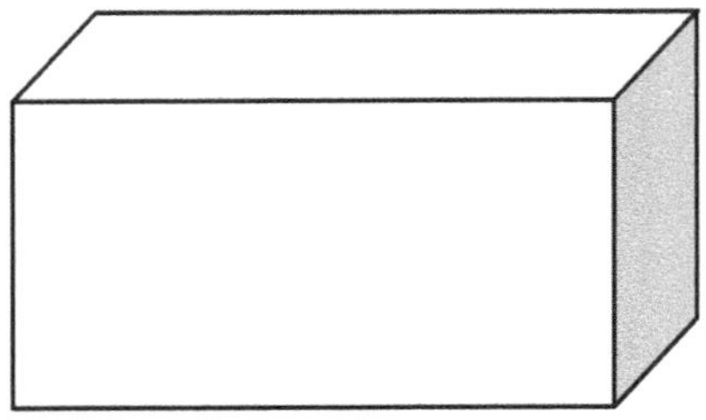

Tick (✔) the shape that matches the shaded face.

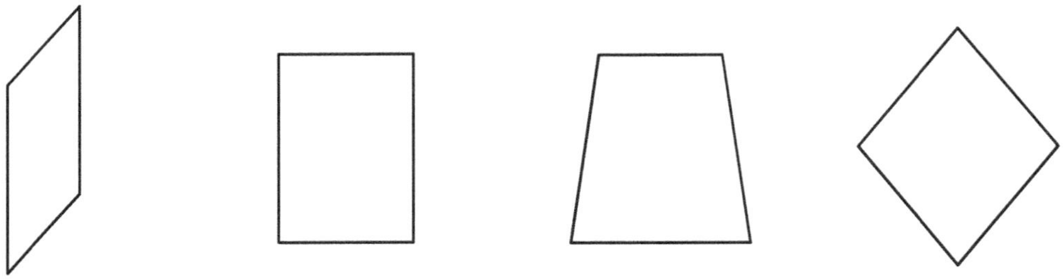

1 mark

2 Write the missing number in the box.

$1\ 701\ 004 = 1\ 000\ 000 + \square + 1000 + 4$

1 mark

3 Write the number six hundred and six thousand and sixty-six in digits.

1 mark

4

drum	teddy	guitar	toy bricks	toy car
£5.75	£3.85	£5.15	£4.25	£4.05

Emma chooses 2 toys from a toy shop.

She spends £9

Which 2 toys does she buy?

☐ and ☐

1 mark

Edward buys a toy car.

How much change does he receive from £5?

☐

1 mark

5 Write the missing numbers in the boxes.

537 639 → (1000 more) → [ ]

1 mark

[ ] → (100 000 more) → 4 019 718

1 mark

6 Here are four lengths.

3 m 7 cm   360 cm   3·4 m $3\frac{1}{2}$ m

Write the lengths in order of size, starting with the smallest.

[ ]	[ ]	[ ]	[ ]
smallest			largest

1 mark

7 Two classes each grew some cress.

They measured the height every week.

The results are shown on the line graph.

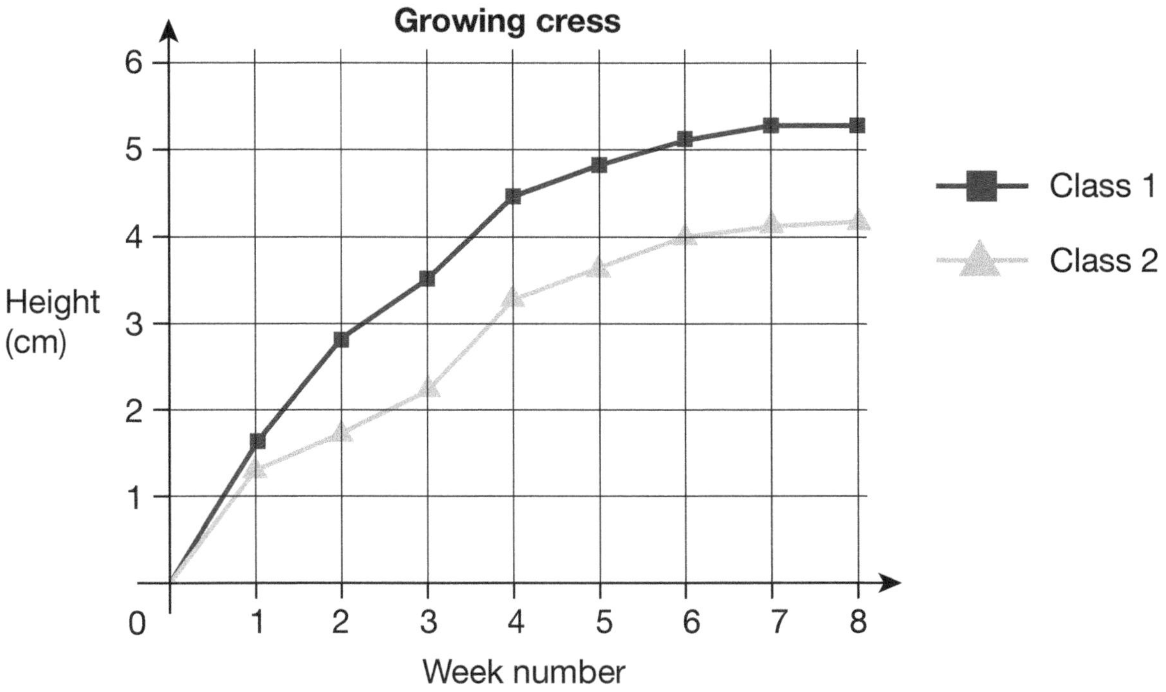

How tall was the cress grown by Class 2 after 6 weeks?

cm

1 mark

After 8 weeks approximately how much taller was the cress grown by Class 1 than Class 2?

cm

1 mark

8 Complete the sequence of the **first 5 prime numbers**.

2				

1 mark

9 Two triangles **A** and **B** are drawn on a square grid.

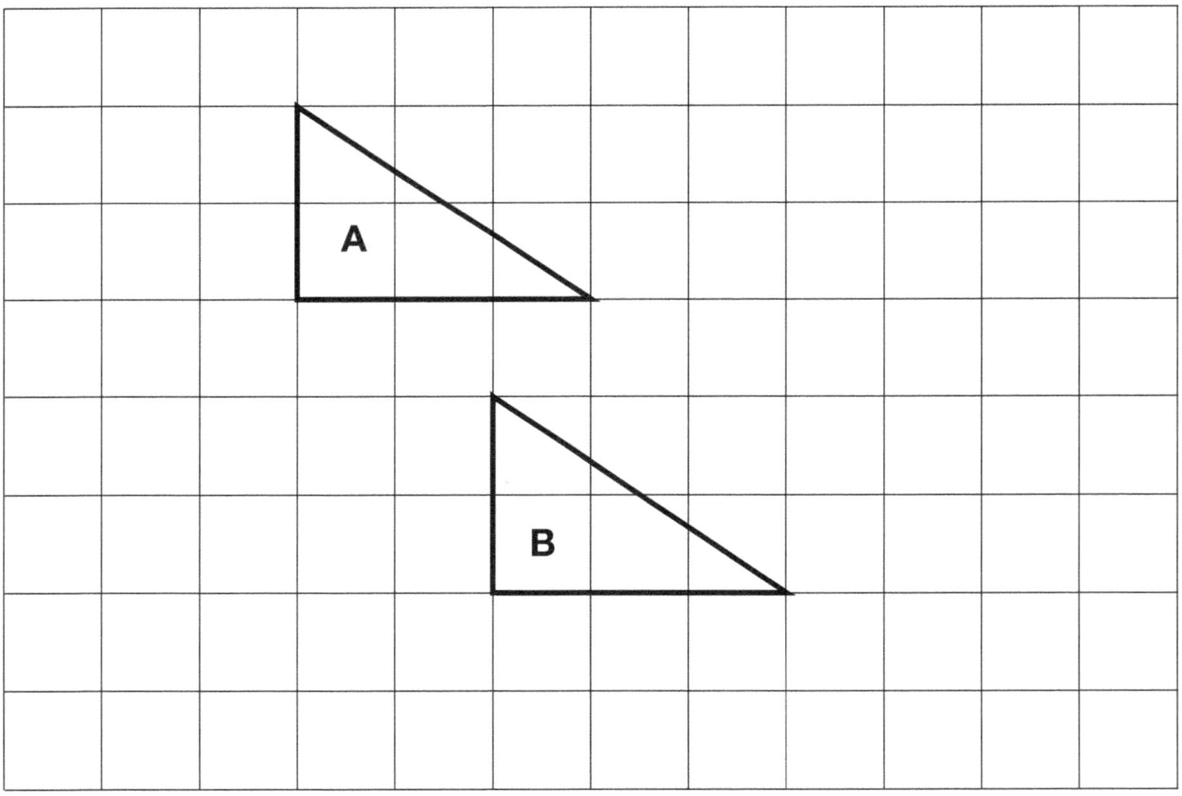

Describe the translation which moves triangle **A** to triangle **B**.

____

1 mark

10 Two function machines are placed next to each other.

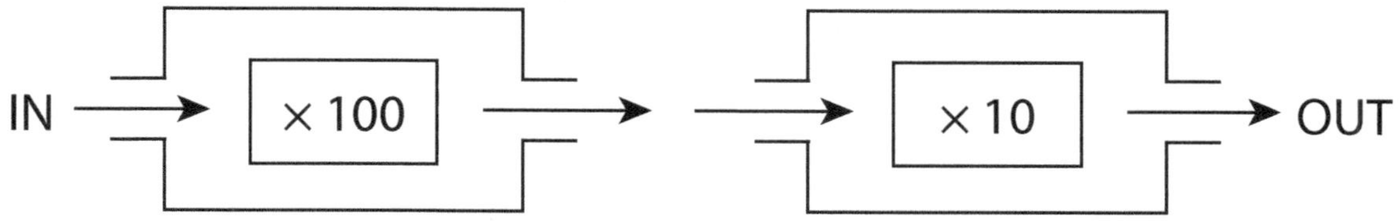

Write the missing number.

In	Out
3	3000
12	12 000
5·4	

1 mark

Write the two functions as a **single function** in this function machine.

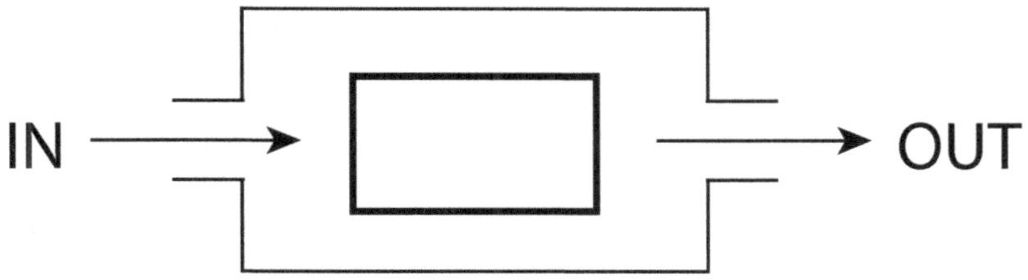

1 mark

11 Join each fraction to its place on the number line.

One has been done for you.

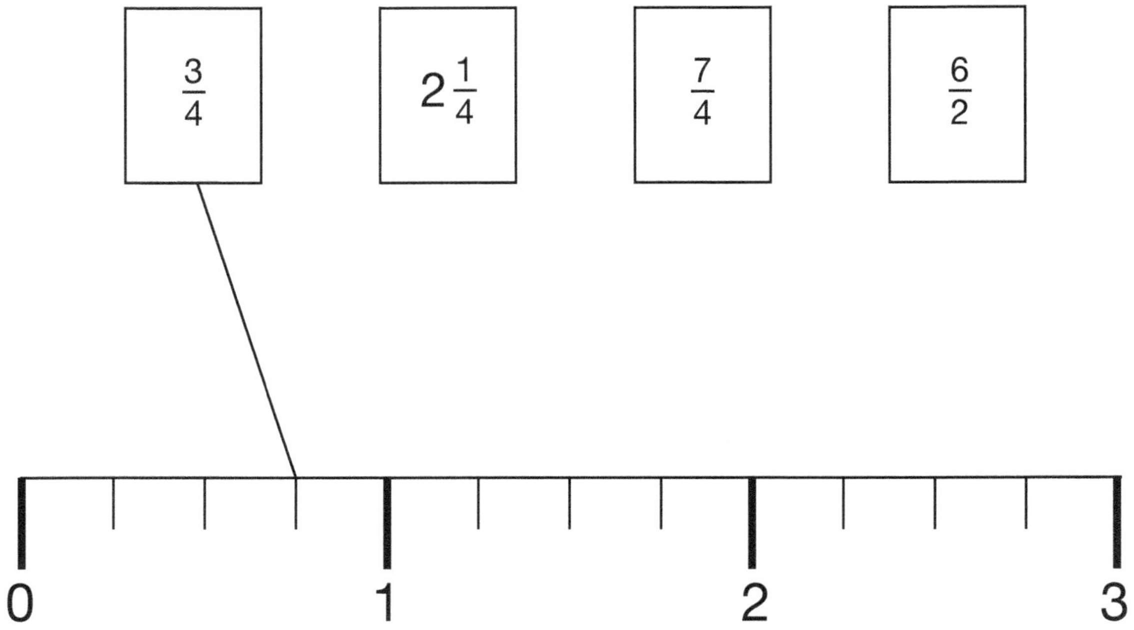

2 marks

12 Harper thinks of a four-digit number.

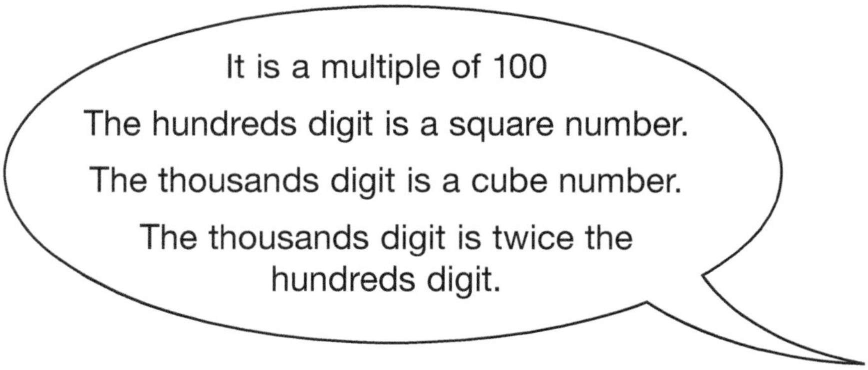

What is the number?

1 mark

13 Write the missing numbers.

	Rounded to a **whole number**	Rounded to **one decimal place**
2·14	2	2·1
5·67		

2 marks

14 A rectangle has a **perimeter of 30 cm**

One side length is 6 cm

Write the missing lengths of the rectangle.

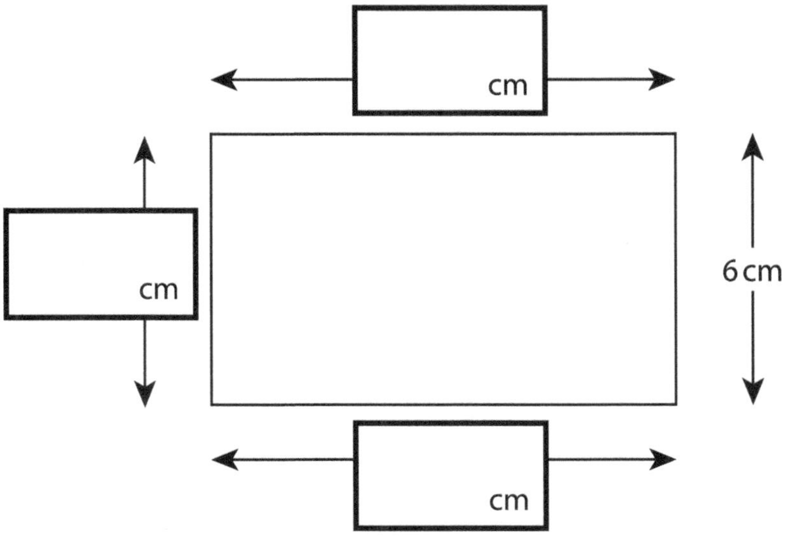

2 marks

15 Match each fraction to a decimal to make a total of 1

One has been done for you.

$\frac{1}{2}$	0·8
$\frac{1}{5}$	0·5
$\frac{3}{4}$	0·25
$\frac{3}{10}$	0·6
$\frac{2}{5}$	0·7

2 marks

16 Write the missing numbers to complete the calculations.

$30 \times 40 = 12 \times \square$

1 mark

$43 \times 12 = 430 + \square$

1 mark

17 A jug of juice holds 1 litre.

The jug can fill 8 mugs exactly.

What is the capacity of one of the mugs?
Give your answer in millilitres.

Show your method

ml

2 marks

18 Write the letters of the shapes that have the **same area**.

Not drawn to scale.

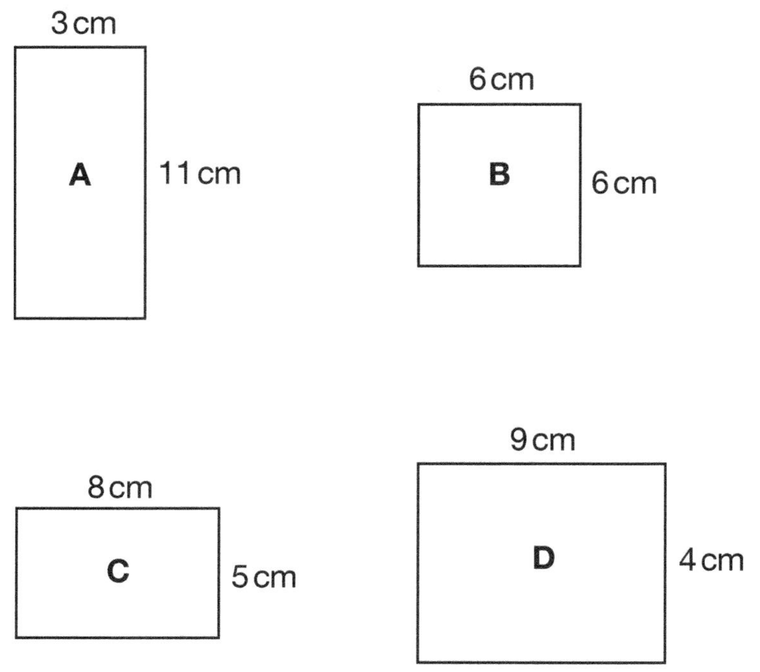

[ ] and [ ]

1 mark

19 8000 people attended a concert, rounded to the nearest thousand.

Circle the number that could **not** be the exact attendance.

8200 7486 8473 7745 8005

1 mark

20 A sheet of stickers has 15 rows and 12 columns.

A teacher has 32 children in the class.

Each child is given 3 stickers each.

How many stickers are left on the sheet?

Show your method

2 marks

21 $\frac{5}{6}$ of the shape is shaded.

Use the diagram to help you find $\frac{5}{6} \times 3$

1 mark

22 Write **seven halves** as a decimal.

1 mark

23 Jamie is designing patterns made from triangles.

He has some triangles that have $\frac{1}{3}$ shaded and some triangles that are all white.

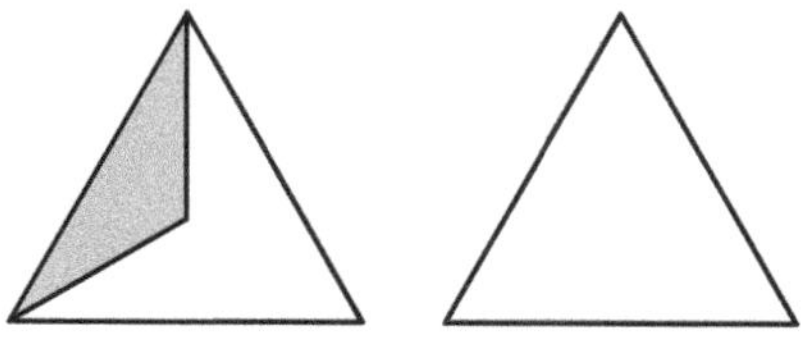

Match each pattern to the fraction it has shaded.

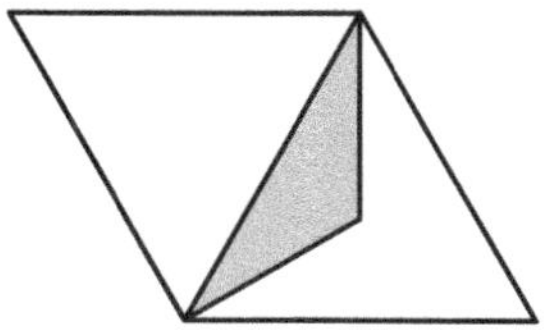

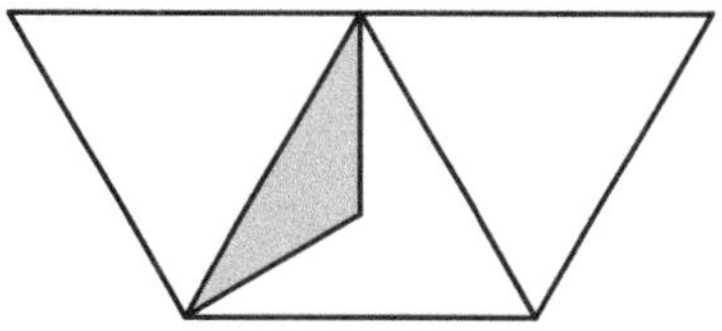

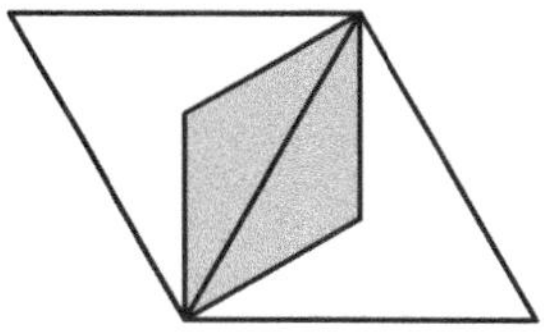

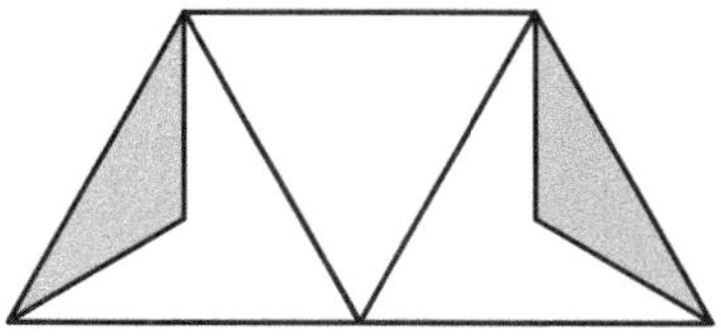

$\frac{1}{9}$

$\frac{1}{3}$

$\frac{2}{9}$

$\frac{1}{6}$

2 marks

Year 5
PAPER
1

Arithmetic

Instructions

You **may not** use a calculator to answer any questions in this test.

Questions and answers

You have **30 minutes** to complete this test.

Work as quickly and as carefully as you can.

Put your answer in the box for each question.

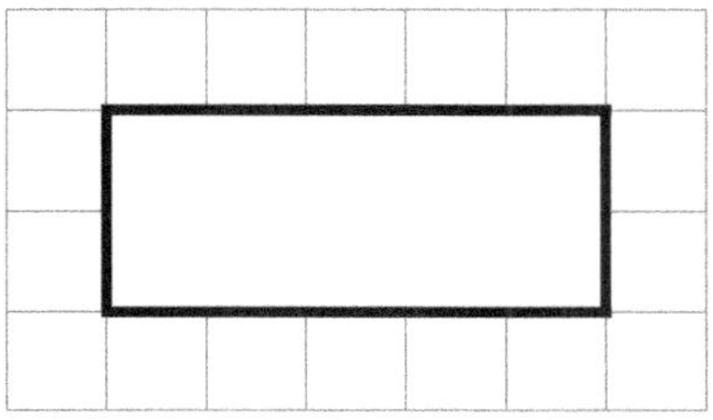

For questions expressed as common fractions, you should give your answer as common fractions.

All other answers should be given as whole numbers or decimals.

If you cannot do one of the questions, **go on to the next one**. You can come back to it later if you have time.

If you finish before the end, **go back and check your work**.

Marks

The number under each box at the side of the page tells you the maximum number of marks for each question.

In this test, some multiplication and division questions are worth 2 marks each. You will be awarded 2 marks for a correct answer. You may get 1 mark for showing a formal method.

All other marks are worth 1 mark each.

1 $18 \times 100 =$

1800

1 mark

2 $10\,000 + 112 =$

10 112

1 mark

3 $\frac{5}{7} - \frac{3}{7} =$

$\frac{2}{7}$

1 mark

4 $11^2 =$

121

1 mark

5 8071 + 3791 =

11 862

1 mark

6 27 000 ÷ 100 =

270

1 mark

7 $1 \cdot 047 + 3 \cdot 109 =$

4·156

1 mark

8 $\frac{1}{2} \times 9 =$

$4\frac{1}{2}$

1 mark

9 £100 – £8.74 =

£91.26

1 mark

10 $213 \times 6 =$

1278

1 mark

11 $13 \times 12 =$

156

1 mark

12 10% of 90 =

9

1 mark

13 $20\,001 + 2001 + 201 =$

22 203

1 mark

14 $0{\cdot}65 + 0{\cdot}35 =$

1

1 mark

15 $168 \div 8 =$

21

1 mark

16 $5 \times 5 \times 5 =$

125

1 mark

17 $\frac{1}{7} + \frac{1}{14} =$

$\frac{3}{14}$

1 mark

18 $40\ 000 - 392 =$

39 608

1 mark

19

$$4\overline{)3\ 5\ 2}$$

Show your method

88

2 marks

20

$\frac{2}{5} + \frac{3}{10} =$

$\frac{7}{10}$

1 mark

21

$3^3 =$

27

1 mark

22 $48 \times 50 =$

2400

1 mark

23 $15\,937 - 8419 =$

7518

1 mark

24

$$\begin{array}{r} 2\,0\,1\,9 \\ \times \quad\quad\quad 3 \\ \hline \end{array}$$

Show your method

6057

2 marks

25 $178 \div 10 =$

17·8

1 mark

26 $14{\cdot}61 + 2{\cdot}571 =$

17·181

1 mark

27 $2 \times 6 \times 50 =$

600

1 mark

28 $99\ 999 + 1000 =$

100 999

1 mark

29 $\frac{4}{5} \times 10 =$

8

1 mark

30 $59 \div 4 =$

14·75

1 mark

31 $300 \div 15 =$

20

1 mark

32

$$\begin{array}{r} 1\ 3\ 2 \\ \times \quad 1\ 7 \\ \hline \end{array}$$

Show your method

2244

2 marks

33 $0{\cdot}07 \times 1000 =$

70

1 mark

34

$7\overline{)1071}$

Show your method

153

2 marks

35

25% of 40 =

10

1 mark

36

$15 \times 1\frac{2}{5} =$

21

1 mark

Year 5
PAPER
2

Reasoning

Instructions

You **may not** use a calculator to answer any questions in this test.

Questions and answers

You have **40 minutes** to complete this test.

Follow the instructions for each question.

Work as quickly and as carefully as you can.

If you need to do working out, you can use the space around the question.

Some questions have a method box like this:

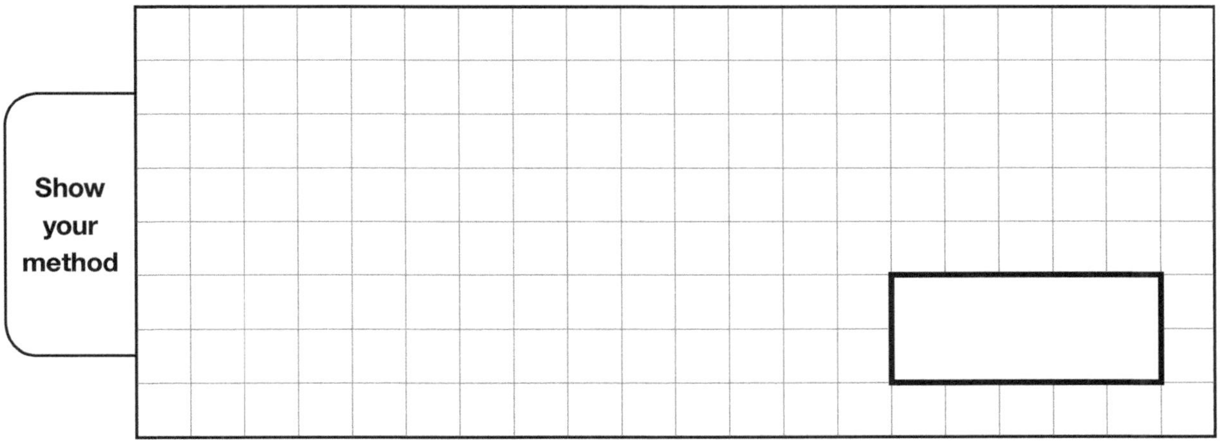

For these questions you may get a mark for showing your method.

If you cannot do one of the questions, **go to the next one.**
You can come back to it later, if you have time.

If you finish before the end, **go back and check your work.**

Marks

The number under each line at the side of the page tells you the maximum number of marks for each question.

1 Circle the number that is between half a million and one million.

80 000   800 000   8 000 000   80 000 000

1 mark

2 The chart shows the distance in kilometres between some cities in Australia.

Alice Springs						
3012	Brisbane					
2324	1717	Cairns				
1511	3415	2727	Darwin			
2270	1674	3054	3781	Melbourne		
3630	4384	5954	4045	3452	Perth	
2644	996	2546	4000	868	4144	Sydney

What is the distance between Brisbane and Darwin?

3415 km

1 mark

3 The rule for a number sequence is to **halve** each time.

Complete the number sequence.

28 → 14 → 7 → 3·5 → 1·75

1 mark

4 Write the numbers in the Carroll diagram.

One has been done for you.

~~16~~ 24 27 30 36

	multiple of 6	**not a multiple of 6**
square number	36	16
not a square number	24, 30	27

2 marks

5 The table shows the minimum temperatures in five different cities on one day.

Boston	7 °C
Chicago	–2 °C
Detroit	4 °C
Buffalo	–6 °C
Denver	–8 °C

The minimum temperature in Denver was 6 degrees colder than in Chicago.

Complete the table.

1 mark

6 Write the missing Roman numeral in the box to complete the sequence.

DXCVIII, DXCIX, DC , DCI

1 mark

7 The table shows the factors of a number, excluding 1 and itself.

Write the missing numbers in the table.

Number	Factors
8	2, 4
28	2, 4, 7, 14
40	2, 4, 5, 8, 10, 20

2 marks

8 Complete the calculation so that the answer is a multiple of 9

55 + [8] = [63]

1 mark

9 Write the fractions and decimal in order of size, starting with the smallest.

$3\frac{3}{5}$ $\frac{15}{5}$ $3{\cdot}5$ $\frac{17}{5}$

$\frac{15}{5}$	$\frac{17}{5}$	$3{\cdot}5$	$3\frac{3}{5}$
smallest			largest

1 mark

10 Write the **cube number** between 50 and 100

64

1 mark

11 Alicia, Kayden and Joe paint a wall for the school play.

Alicia paints $\frac{2}{9}$ of the wall.

Kayden paints $\frac{1}{3}$ of the wall.

What fraction of the wall does Joe paint?

Show your method

$\frac{4}{9}$

2 marks

12 Write the missing number in the box.

$$13\ 401 + \boxed{36\ 599} = 50\ 000$$

1 mark

13 One crate holds 6 bottles.

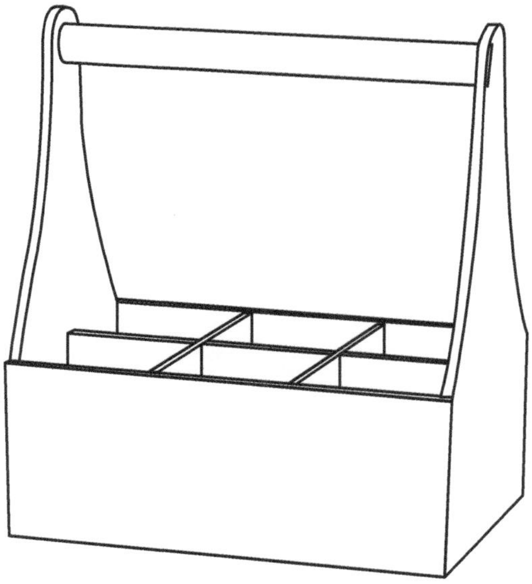

How many crates are needed to hold 204 bottles?

34

1 mark

14 Calculate the size of angle *a*

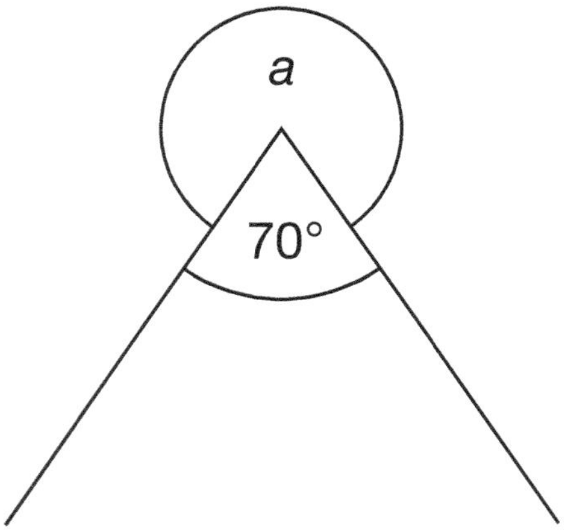

290 °

1 mark

Complete the table.

	type of angle
angle ***a***	reflex
angle of 70°	acute

1 mark

15 Write 0·47 as a fraction.

$\frac{47}{100}$

1 mark

16 A pond is approximately 3 feet deep.

Tick (✔) the measure that is approximately 3 feet in metric units.

9 cm ☐

30 cm ☐

90 cm ☑

3 m ☐

9 m ☐

1 mark

17 **Complete the table.**

The first row has been done for you.

	Less than $\frac{1}{2}$	Greater than $\frac{1}{2}$
$\frac{3}{4}$		✔
$\frac{11}{20}$		✔
$\frac{42}{100}$	✔	

1 mark

18 The mass of an orange is 130 g

A pineapple is **8 times heavier**.

Give units with your answer.

Calculate the total mass of the orange and the pineapple.

Show your method

1170 g
or 1.170 kg

2 marks

19 Here are five cards.

0	1	2	3	·

Use each card once to make the largest number less than 1

0	·	3	2	1

1 mark

20 A regular pentagon is cut to create a trapezium and a triangle.

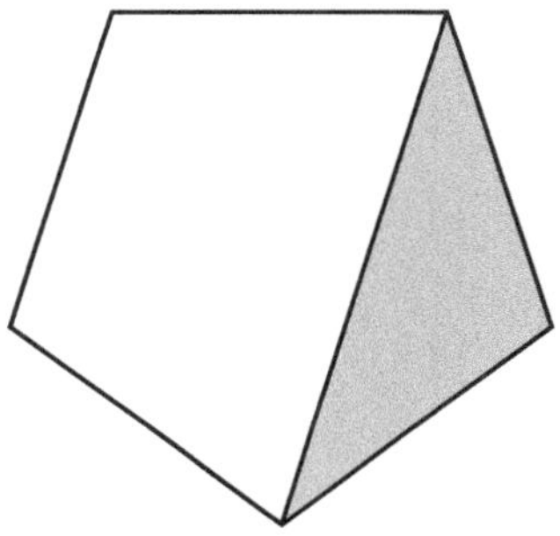

Tick (✔) the type of triangle.

isosceles ☑ equilateral ☐ scalene ☐

Explain your answer.

For example, the pentagon is regular and so two sides of the triangle are equal.

1 mark

21 Here is a number fact:

$$70 \times 30 = 2100$$

Draw a line to join **two** cards that have a total of 2100

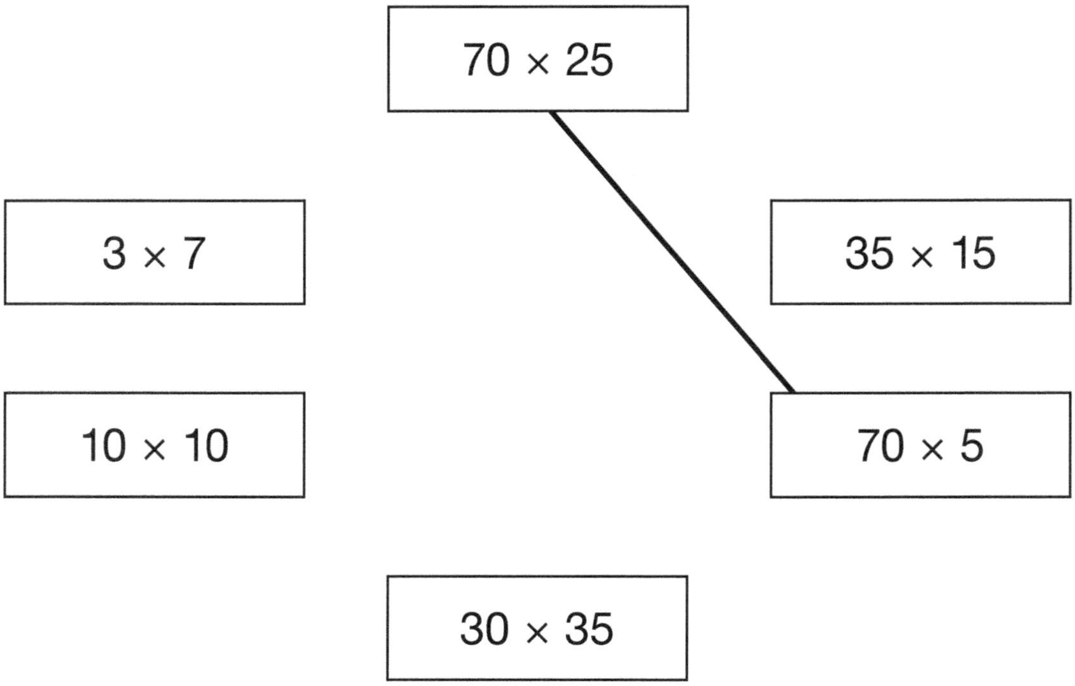

1 mark

22 Here is a bus timetable for the town of Thriplow.

Bus timetable

Station	0917	1107	1314	1515	1714
Park	0934	1124	1331	1552	1751
Market	1015	1205	1412	1613	1812
Library	1057	1247	1454	1655	1854
Airport	1137	1327	1534	1735	1934

Jake is at the park at a **quarter to eleven**.

What time does the next bus depart?

1124

1 mark

The bus takes 2 hours and 3 minutes to travel from the park to the airport.

Write this time in minutes.

123 minutes

1 mark

23 Here are six calculations.

Tick (✔) the calculation that gives the **largest** answer.

703 – 99	☐	5·2 × 104	☐
1197 ÷ 3	☐	5000 ÷ 504	☐
604 + 97	☑	32 × 9	☐

1 mark

24 Find the perimeter of the shape.

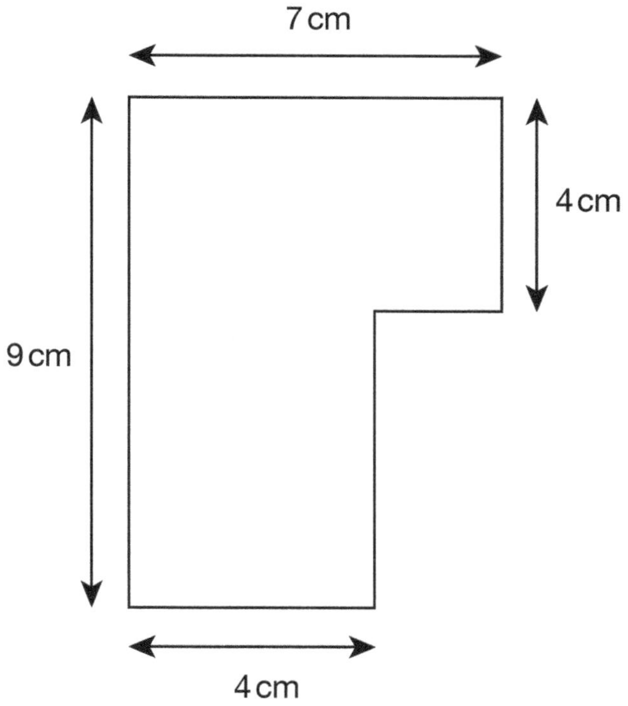

Show your method

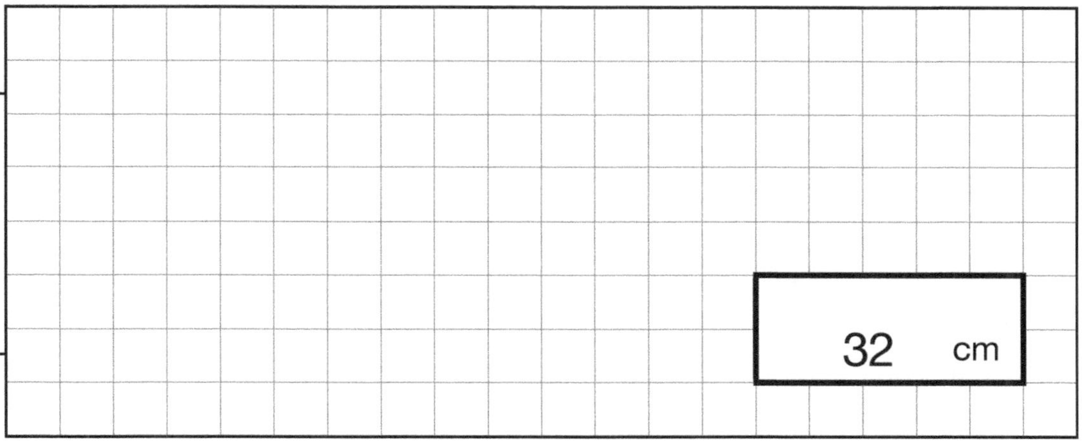

32 cm

2 marks

25 Francesca says:

Show that she is wrong, using an example.

For example, 9 x 7 or 7 is a factor.

1 mark

26 The multiples of 5 are shaded on the 100 square.

1	2	3	4	5	6	7	8	9	10
11	12	13	14	15	16	17	18	19	20
21	22	23	24	25	26	27	28	29	30
31	32	33	34	35	36	37	38	39	40
41	42	43	44	45	46	47	48	49	50
51	52	53	54	55	56	57	58	59	60
61	62	63	64	65	66	67	68	69	70
71	72	73	74	75	76	77	78	79	80
81	82	83	84	85	86	87	88	89	90
91	92	93	94	95	96	97	98	99	100

What **percentage** of the numbers from 1 to 100 are multiples of 5?

20 %

1 mark

27 A piece of material measures 6·5 m by 1 m

How many 50 cm squares can be cut from the material?

Show your method

26

2 marks

Year 5
PAPER
3

Reasoning

Instructions

You **may not** use a calculator to answer any questions in this test.

Questions and answers

You have **40 minutes** to complete this test.

Follow the instructions for each question.

Work as quickly and as carefully as you can.

If you need to do working out, you can use the space around the question.

Some questions have a method box like this:

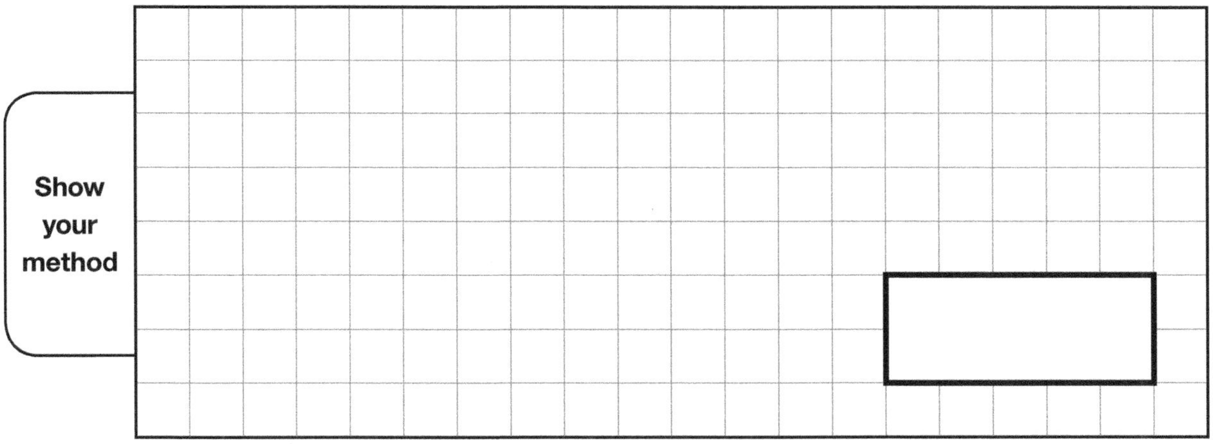

For these questions you may get a mark for showing your method.

If you cannot do one of the questions, **go to the next one.**
You can come back to it later, if you have time.

If you finish before the end, **go back and check your work.**

Marks

The number under each line at the side of the page tells you the maximum number of marks for each question.

1 One face of the 3-D shape is shaded.

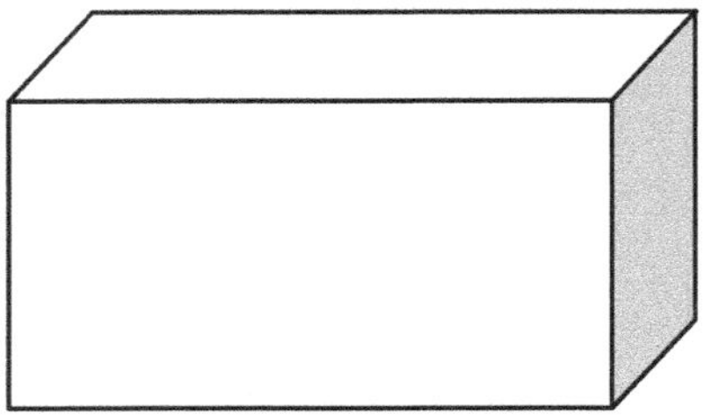

Tick (✔) the shape that matches the shaded face.

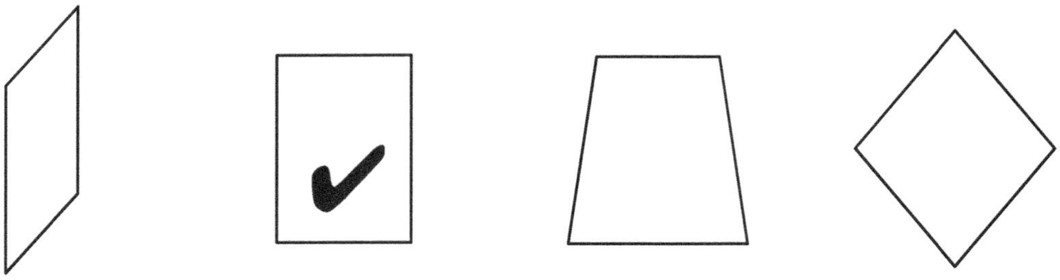

1 mark

2 Write the missing number in the box.

1 701 004 = 1 000 000 + | 700 000 | + 1000 + 4

1 mark

3 Write the number six hundred and six thousand and sixty-six in digits.

606 066

1 mark

4

drum	teddy	guitar	toy bricks	toy car
£5.75	£3.85	£5.15	£4.25	£4.05

Emma chooses 2 toys from a toy shop.

She spends £9

Which 2 toys does she buy?

teddy and guitar

1 mark

Edward buys a toy car.

How much change does he receive from £5?

95p or £0.95

1 mark

5 Write the missing numbers in the boxes.

537 639 → (1000 more) → 538 639

1 mark

3 919 718 → (100 000 more) → 4 019 718

1 mark

6 Here are four lengths.

3 m 7 cm   360 cm   3·4 m $3\frac{1}{2}$ m

Write the lengths in order of size, starting with the smallest.

3 m 7 cm	3·4 m	$3\frac{1}{2}$ m	360 cm
smallest			largest

1 mark

7 Two classes each grew some cress.

They measured the height every week.

The results are shown on the line graph.

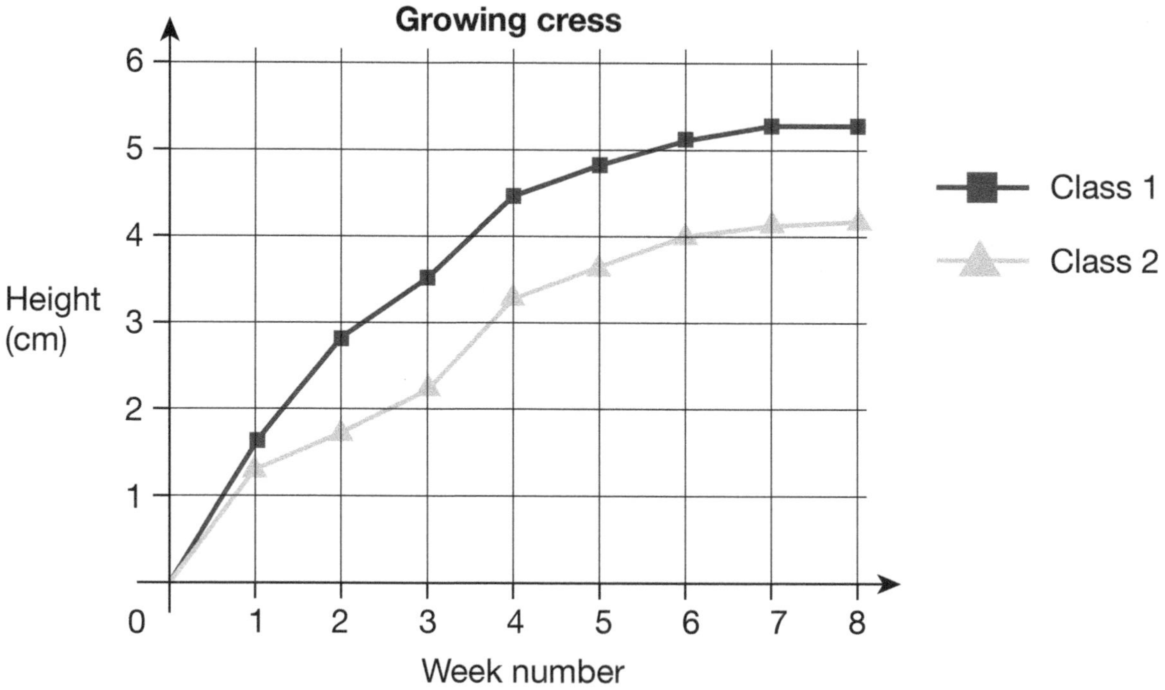

How tall was the cress grown by Class 2 after 6 weeks?

4 cm

1 mark

After 8 weeks approximately how much taller was the cress grown by Class 1 than Class 2?

1 cm

1 mark

8 Complete the sequence of the **first 5 prime numbers**.

2	3	5	7	11

1 mark

9 Two triangles **A** and **B** are drawn on a square grid.

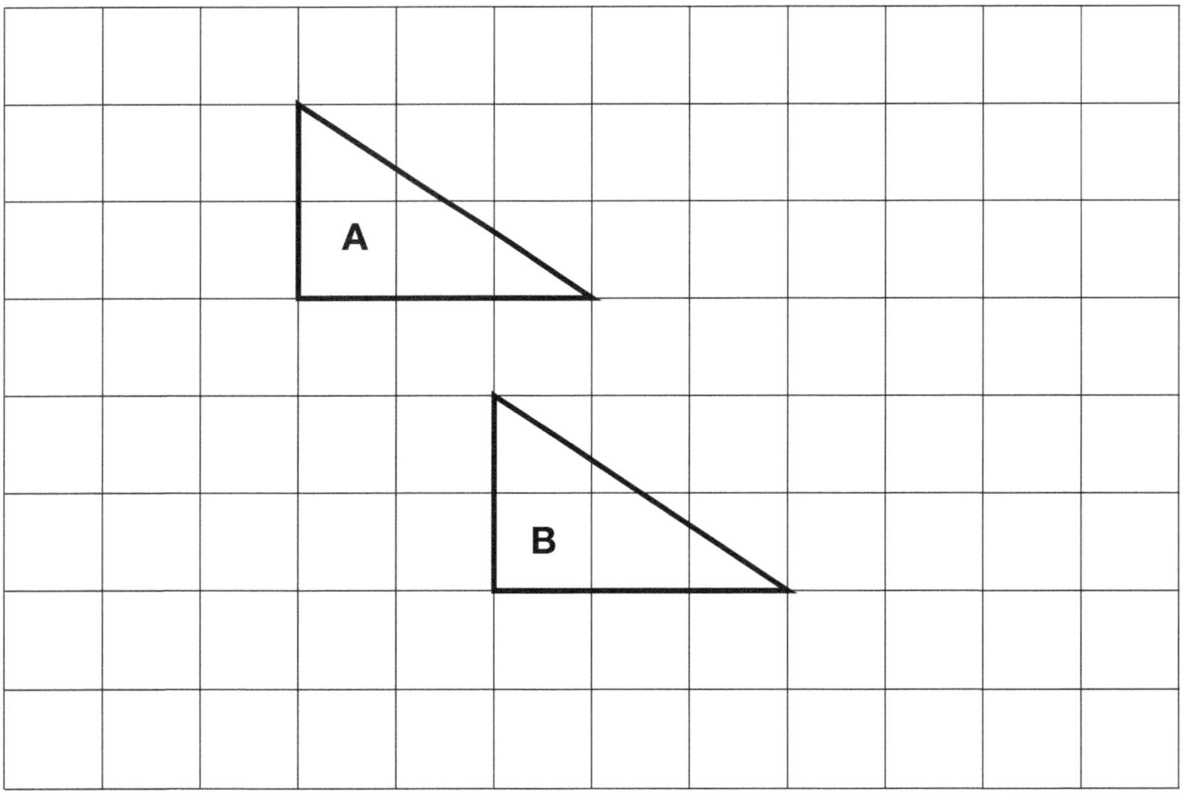

Describe the translation which moves triangle **A** to triangle **B**.

RIGHT 2 DOWN 3

1 mark

10 Two function machines are placed next to each other.

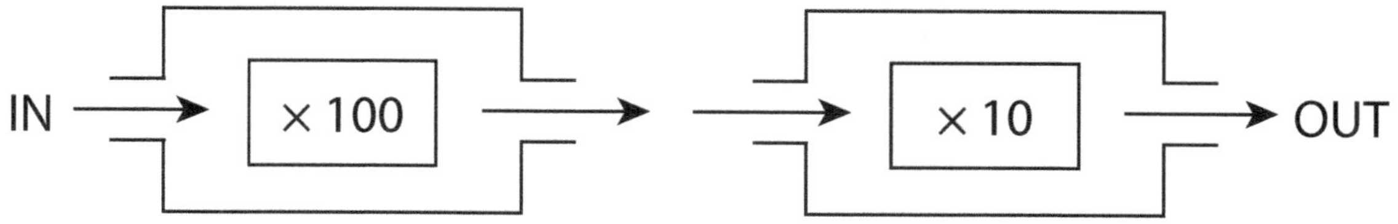

Write the missing number.

In	Out
3	3000
12	12 000
5·4	5400

1 mark

Write the two functions as a **single function** in this function machine.

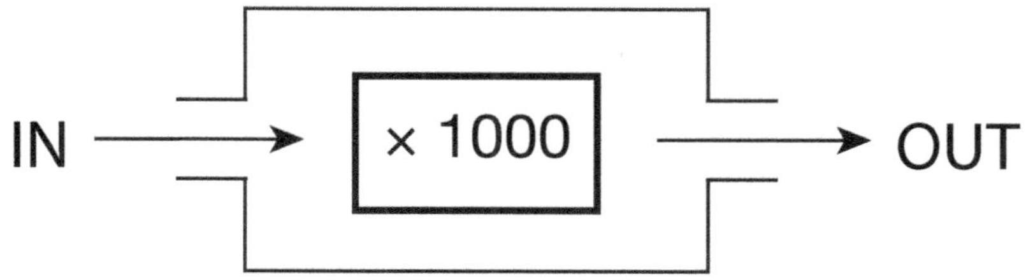

1 mark

11 Join each fraction to its place on the number line.

One has been done for you.

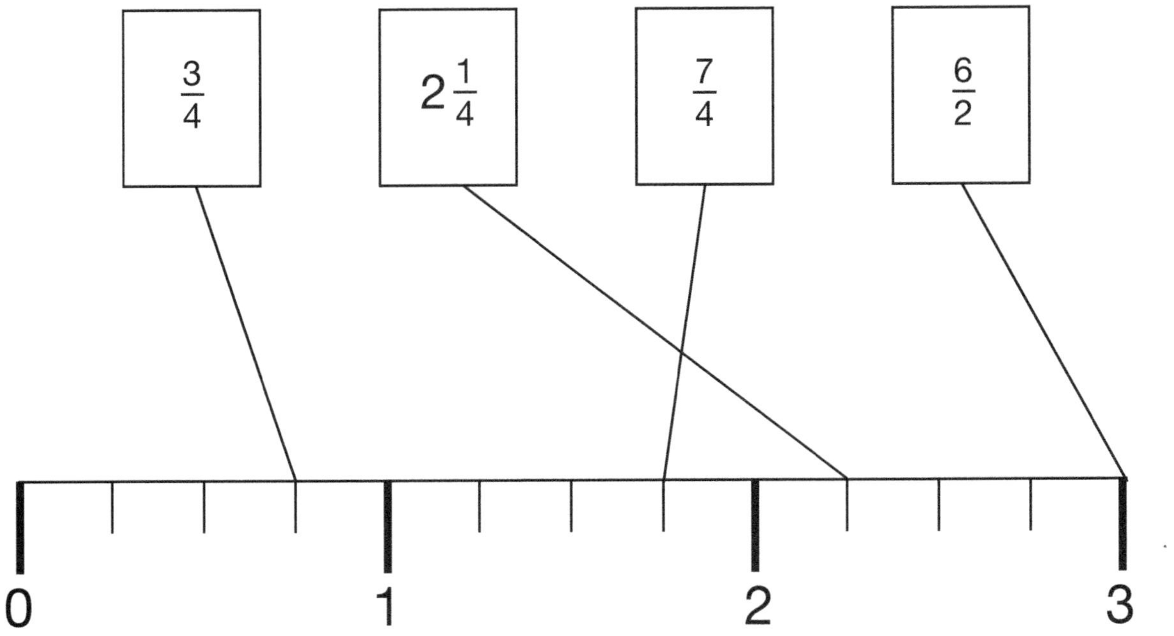

2 marks

12 Harper thinks of a four-digit number.

It is a multiple of 100

The hundreds digit is a square number.

The thousands digit is a cube number.

The thousands digit is twice the hundreds digit.

What is the number?

8	4	0	0

1 mark

13 Write the missing numbers.

	Rounded to a **whole number**	Rounded to **one decimal place**
2·14	2	2·1
5·67	6	5·7

2 marks

14 A rectangle has a **perimeter of 30 cm**

One side length is 6 cm

Write the missing lengths of the rectangle.

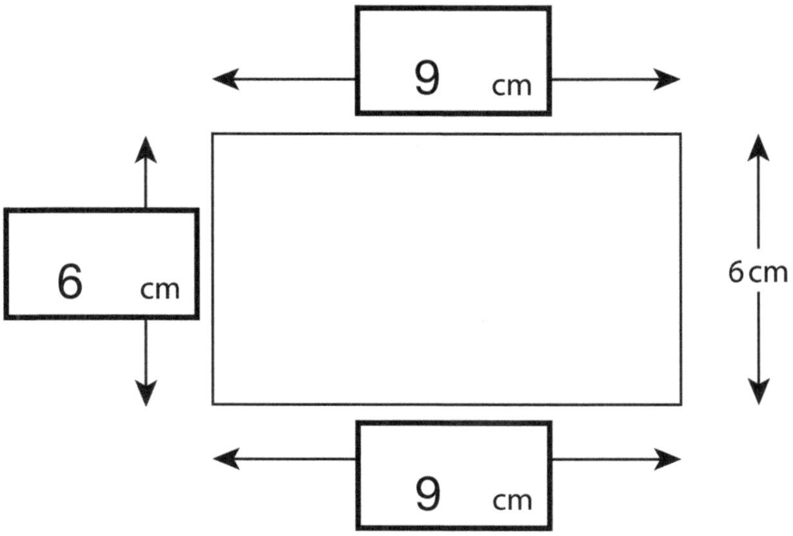

2 marks

15 Match each fraction to a decimal to make a total of 1

One has been done for you.

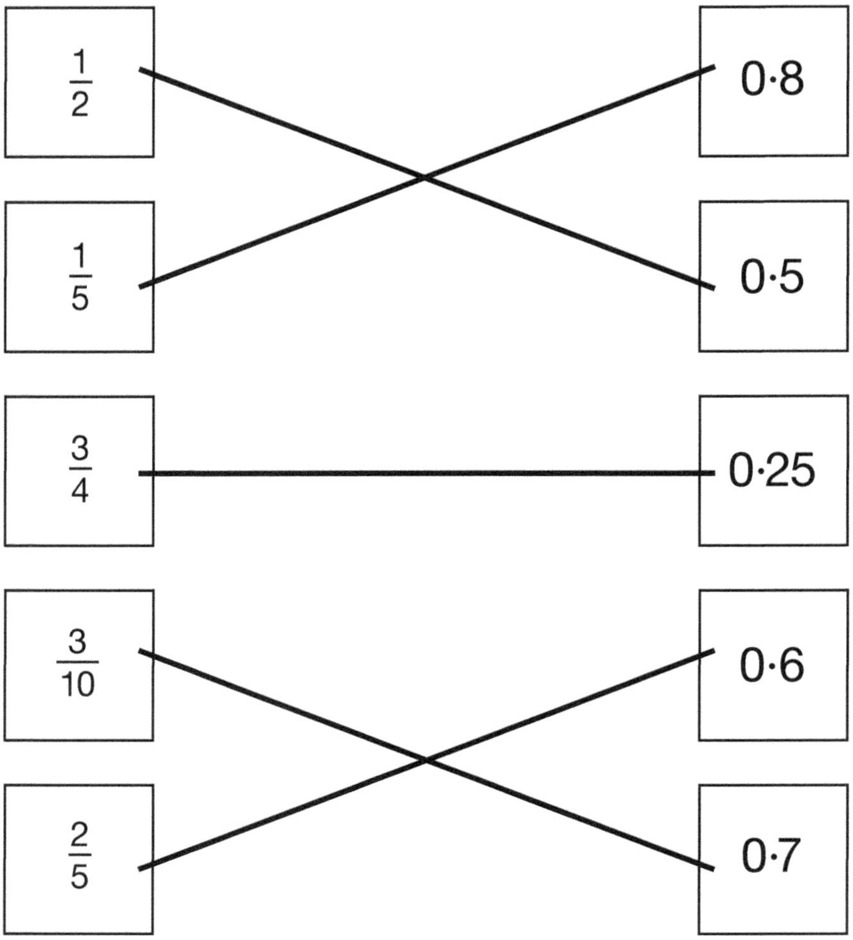

2 marks

16 Write the missing numbers to complete the calculations.

$30 \times 40 = 12 \times$ [100]

1 mark

$43 \times 12 = 430 +$ [86]

1 mark

17 A jug of juice holds 1 litre.

The jug can fill 8 mugs exactly.

What is the capacity of one of the mugs?
Give your answer in millilitres.

Show your method

125 ml

2 marks

18 Write the letters of the shapes that have the **same area**.

Not drawn to scale.

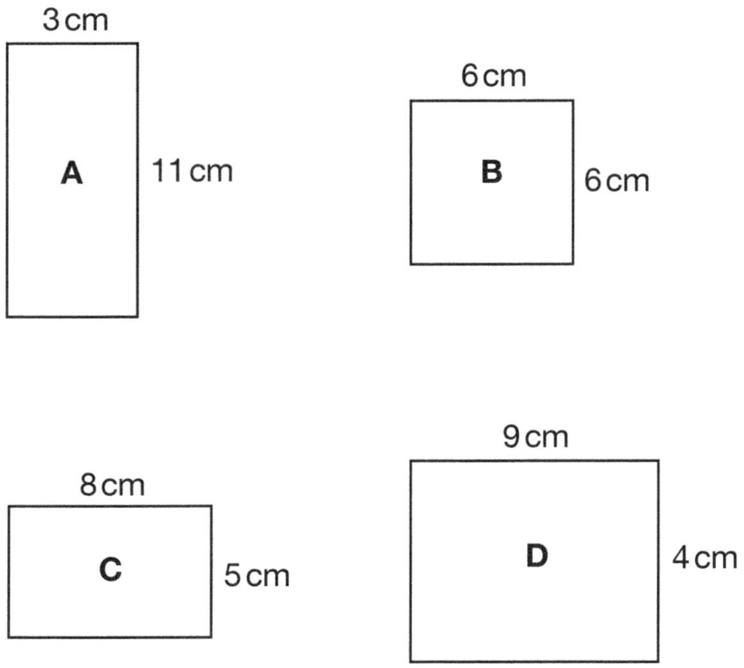

B and D

1 mark

19 8000 people attended a concert, rounded to the nearest thousand.

Circle the number that could **not** be the exact attendance.

8200 7486 8473 7745 8005

1 mark

20 A sheet of stickers has 15 rows and 12 columns.

A teacher has 32 children in the class.

Each child is given 3 stickers each.

How many stickers are left on the sheet?

Show your method

84

2 marks

21 $\frac{5}{6}$ of the shape is shaded.

Use the diagram to help you find $\frac{5}{6} \times 3$

$\frac{15}{6}$

1 mark

22 Write **seven halves** as a decimal.

3·5

1 mark

23 Jamie is designing patterns made from triangles.

He has some triangles that have $\frac{1}{3}$ shaded and some triangles that are all white.

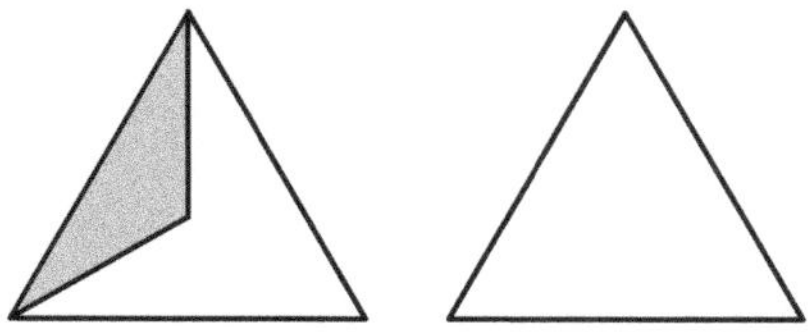

Match each pattern to the fraction it has shaded.

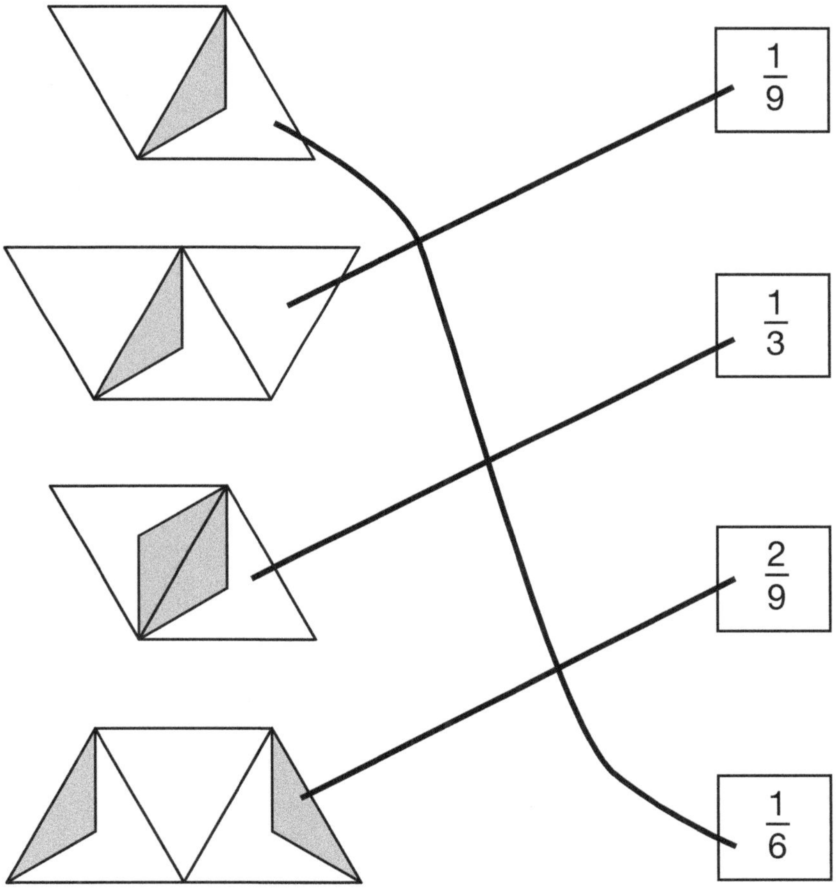

2 marks